Gabriel Castellano
Didier Gastmans

Carbon dioxide emissions from the soil in the Atlantic Forest

AF144335

Gabriel Castellano
Didier Gastmans

Carbon dioxide emissions from the soil in the Atlantic Forest

Semideciduous Seasonal Forest

ScienciaScripts

Imprint

Any brand names and product names mentioned in this book are subject to trademark, brand or patent protection and are trademarks or registered trademarks of their respective holders. The use of brand names, product names, common names, trade names, product descriptions etc. even without a particular marking in this work is in no way to be construed to mean that such names may be regarded as unrestricted in respect of trademark and brand protection legislation and could thus be used by anyone.

Cover image: www.ingimage.com

This book is a translation from the original published under ISBN 978-620-2-04929-0.

Publisher:
Sciencia Scripts
is a trademark of
Dodo Books Indian Ocean Ltd. and OmniScriptum S.R.L publishing group

120 High Road, East Finchley, London, N2 9ED, United Kingdom
Str. Armeneasca 28/1, office 1, Chisinau MD-2012, Republic of Moldova, Europe
Printed at: see last page
ISBN: 978-620-7-24418-8

SUMMARY

1. INTRODUCTION

Emissions of greenhouse gases (CO_2, CH_4, N_2O and others present in the atmosphere) have become one of the main environmental concerns of our time (KUNTORO, 2009). Among these gases, carbon dioxide (CO_2) is responsible for around 60% of the intensification of the greenhouse effect (FERNANDES, 2003), since since the beginning of the industrial revolution the concentrations of this gas in the atmosphere have risen from 280 ppm to around 390 ppm (DENMAN et al., 2007).

One of the main causes of the increase in CO_2 concentrations in the atmosphere is related to the intensification of anthropogenic activities, such as changes in land use and land cover, i.e. the replacement of native biomes by cutting and burning the vegetation that has been removed, promoting the replacement of local plant species and communities with agricultural activities for economic purposes. It is estimated that these changes in land use, which occur preferentially in savannah and forest environments because the soil and climate conditions of these biomes are ideal for high-yield agricultural production, are responsible for around 30% of total CO_2 emissions into the atmosphere (SABINE et al., 2004).

Emissions of carbon dioxide under these conditions are caused both by the burning of native vegetation and by conventional agriculture, which is less efficient at accumulating organic and microbial carbon in the soil than areas planted with conservation agriculture or forest (CARDOSO et al., 2010).

During the 1980s and 1990s, emissions caused by deforestation and the removal of forest biomass were estimated at around 10^9 tons of carbon per year (WATSON et al., 2000). If the predicted climate changes materialize, the impacts on forests will be profound and long-lasting, varying from region to region, affecting both the distribution and composition of forests (IPCC, 2001; FAO, 2001).

In this context, new demands for research into forest restoration have arisen, especially those related to quantifying the environmental services provided by reforestation with native species and discussing the effectiveness of this strategy in reducing atmospheric CO_2 levels (FOSTER E MELLO, 2007).

Because tropical ecosystems (soil and vegetation) account for between 20 and 25% of the world's terrestrial carbon, associated with their enormous stock of carbon stored in the soil (SCHLESINGER, 1997) and their role in the biogeochemical processes that lead to the regulation of global warming (FERNANDES, 2003), studies on the dynamics of this element in the soil are being highlighted, as well as modeling of climate change. In this context, Kutsch et al. (2010) present some questions regarding the capacity of ecosystems to sequester CO_2, such as:

1)   How much CO_2 can the soil sequester in each ecosystem on the globe? And how long does this carbon remain in the soil?

2)   Will the increase in the net primary production of the ecosystem, due to the increase in the concentration of atmospheric CO_2, associated with anthropogenic action, such as nitrogen fertilization, increase the production of litter, and consequently increase the carbon stock in the soils?

Forest biomes are efficient carbon stores, with forests holding approximately half of the total carbon stored by terrestrial vegetation. Boreal forests account for 26% of total terrestrial carbon stocks, while tropical and temperate forests contain 20% and 7%, respectively (DIXON et al., 1994).

Brazil is the fifth largest country in terms of land area, with approximately 5.7% of the planet's land surface and 47.3% of the area of South America. It also has an impressive natural heritage, which places it at the top of the list of megadiverse countries, those with the greatest number of plant and animal species (CAMPANILI E SCHAFFER, 2010).

Among the main Brazilian biomes, the Atlantic Rainforest, which originally covered an area of 1,300,000 km^2 , stretching across 17 Brazilian states, today has only 27% of its original territory. It is made up of a set of forest formations, as well as associated ecosystems: natural grasslands, restingas and mangroves, with remnants distributed in thousands of vegetation fragments, which still retain high levels of fauna and flora and provide invaluable environmental services by protecting water sources, containing slopes and regulating the climate (CAMPANILI AND SCHAFFER, 2010).

The Semideciduous Seasonal Forest is one of the most degraded and fragmented formations of the Atlantic Forest in the state of Sao Paulo, as it is located in regions of Brazil that have undergone major economic transformations dependent on agricultural and livestock production processes. Genera of Amazonian origin dominate this forest, including: *Parapiptadenia, Peltophorum, Cariniana, Lecythis, Tabebuia* and *Astronium* (VELOSO et al., 1991). The tree formations that cover eutrophic basaltic soils are rare to find, as the soil is highly valued for agricultural production.

The removal of wood from the formations of the Semideciduous Seasonal Forest, especially in the upper stratum, was so extensive, especially during the 20th century, that today it is doubtful whether there are any remnants that have not suffered strong anthropic pressures in the past (RODRIGUES, 1999).

Therefore, the recovery of the Atlantic Forest plays an important role as an important ecosystem regulator of CO_2, and not just in relation to biodiversity and other related attributes, which is why the Atlantic Forest Restoration Pact was created. According to the protocol established in this pact, 15

3

million hectares are to be planted and restored throughout Brazil by the year 2050, distributed in annual plans. This process will bring about a regional change in land use and occupation, which should alter CO_2 balances regionally and globally. Among the protocol's priorities is the valuation of the environmental or ecosystem services offered to society by the remaining areas and those under restoration, reinforcing their importance for the quality of life and the means of production, taking advantage of opportunities in the carbon and water markets.

However, in order for these services to be properly valued, a broad study of the biogeochemical cycles of carbon in the Atlantic Forest is needed, making it a priority to evaluate and characterize CO_2 emissions in areas with different types of soils and forest physiognomies in this biome, since these emissions can be an important indicator of the environmental quality of the soil, as well as guiding planting and restoration plans.

1.1 Objectives

The main objective of this study was to characterize soil CO_2 emission rates in two native forest areas within the Atlantic Forest morphoclimatic domain, located in the Edmundo Navarro de Andrade State Forest (FEENA), planted in 1918 and 2014. Secondary objectives include:

• Correlate these emissions with atmospheric and soil physico-chemical parameters: pressure, air temperature, air humidity, air temperature, air humidity, thermal resistivity coefficient, soil carbon content and C/N ratio;

• Build a robust statistical model based on the observed correlations, capable of predicting emission rates for the area studied.

• To evaluate the functioning, under field conditions, of the flow chamber operating system coupled to the infrared gas meter developed by Moreno (2012).

4

2. LITERATURE review

2.1 Biogeochemical Carbon Cycle

Carbon is an essential element for life on the planet, a constituent of the organic molecules and tissues of living organisms. It is incorporated from the atmosphere by plants through photosynthesis to form glucose ($C_6H_nO_6$), the constituent of organic matter. It is returned to the atmosphere by the respiration of producer, consumer and decomposer organisms (CALIJURI, 2013).

One of its main forms of occurrence is in combination with oxygen, forming carbon dioxide molecules, which are present in the atmosphere (the largest reservoir), or dissolved in the waters of seas, rivers and lakes, or even incorporated into the soil in the form of organic matter (DIAS, 2006).

The carbon cycle has been altered by anthropogenic activity in recent years, whether through the burning of fossil fuels, changes in land use and occupation through the cutting down of forests and the burning of biomass, or volcanic activity. It is estimated that anthropogenic activities currently contribute seven billion tons of CO_2 to the atmosphere every year. Half of this carbon remains in the atmosphere, and the rest is dissolved in the oceans or sequestered by photosynthetic activity, retained in biomass or added to soil organic matter (SCHLESINGER, 1997; GRACE, 2001).

In the aquatic environment, atmospheric CO_2 combines with water through diffusion to form carbonic acid (H_2CO_3), which is rapidly dissociated into H ions$^+$, bicarbonate (HCO_3^{-1}) and carbonate (CO_3^{-2}), according to the following reaction:

$$CO_2 + H_2O \quad \leftrightarrow \quad H_2CO_3 \quad \leftrightarrow \quad H^+ + HCO_3^{-1} \quad \leftrightarrow \quad 2H^+ + CO_3^{-2} \quad (1)$$

This reaction is reversible and always occurs in the direction of the component with the highest concentration towards the one with the lowest concentration, both in water and in the air, i.e. the reaction indicates that when there is an increase in the concentration of CO_2 in the atmosphere, the oceans will absorb more CO_2, which will remain dissolved in the water in the form of bicarbonate or carbonate (CALIJURI, 2013).

If calcium ions are available in the water, they can also react with carbonate and bicarbonate ions to form calcium carbonate, which will precipitate due to its low solubility, accumulating in sediments, according to the reaction below:

$$Ca^{+2} + CO_3^{-2} \quad \rightarrow \quad CaCO_3 \, (2)$$

Under acidic pH conditions, the formation of carbonic acid removes carbon from the system. This removal reduces the amount of $CaCO_3$, which in turn increases the rate of dissolution of calcareous rocks. When these slightly acidic calcium-laden waters meet the higher pH waters of the ocean, the $CaCO3$ can precipitate again and accumulate in the sediment (CALIJURI, 2013).

In the marine environment, under neutral conditions, the carbon system remains in equilibrium, according to the reaction below:

The activity of organisms can affect this reaction. The removal of CO_2 by photosynthesis shifts the balance to the left, favoring the formation and precipitation of calcium carbonate (CALIJURI, 2013).

In continental areas, the largest carbon reservoir is represented by soils, which store around 40×10^{18} g of carbon, while vegetation cover has an estimated carbon stock of 56×10^{16} g (SCHLESINGER, 1997; GRACE, 2001). Tropical forest soils act as a source and sink of various gases, including CO_2, and play a significant role in the physico-chemical processes of the atmosphere (KELLER et al., 1986).

Through photosynthesis, it is estimated that every year around 60×10^{1} 5 g of carbon is fixed in plant tissues, and almost all of it returns to the atmosphere through respiration of living tissues and the soil (SCHLESINGER, 1997). The natural and cyclical processes known as the carbon cycle comprise photosynthesis, respiration and dissolution (Figure 1).

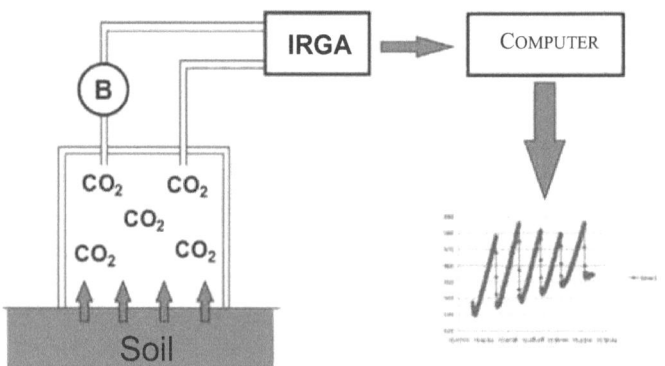

FIGURE 1. MAIN ANNUAL CARBON STOCKS AND FLOWS (IN PGC). SOURCE: ADAPTED FROM SCHLESINGER (1997) BY DIAS (2006).

2.2 Carbon Stocks and Fixation in Tropical Soils

The dynamic quantities of humus, or carbon in the soil, are determined by the set of soil and climatic factors and the management of the soil-plant system, which control the rates of deposition, incorporation and decomposition of carbon in the soil (SIQUEIRA E FRANCO, 1988). In a soil in

6

equilibrium with vegetation, the carbon content (C) is given by the formula:

$C = A/K$, where $A = b. M$ (4)

Where: **C**, represents the content (%) or quantity (t.ha^{-1}) of carbon in the soil which, if multiplied by the value of 1.724, corresponds to the soil's organic matter (OM); **A** is the annual addition of carbon to the soil (t.ha^{-1}); **K** represents the annual rate of decomposition of the soil's organic carbon; **b** is the quantity (t.ha^{-1}) of OM-fresh organic matter (dead branches, leaves and roots) and **m** is the conversion rate.

When areas are restored with native forests, plant residues are added to the soil, resulting in an accumulation of carbon. Long-term experiments show that there is a positive linear relationship between the input of plant residues (BAYER, 1996; LOVATO et al., 2004), or other carbon sources (NICOLOSO, 2009), with the increase in carbon concentrations in the surface centimeters of the soil in agricultural areas, showing that cultivated tropical and subtropical soils are efficient carbon accumulators (Figure 2).

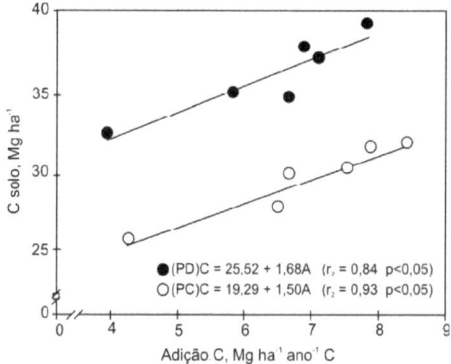

Figure 2 - Relationship between carbon input by agricultural systems in Argissolo subjected to Direct Planting (PD) and Conventional Planting (PC) SOURCE: Bayer et al. (2011).

Organic matter has a high cation exchange capacity (CEC), which varies from 300 to 1400 meq.100g^{-1} , as well as having a buffering effect on the soil, which is related to the soil's ability to maintain its pH unchanged when treated with acid (fertilization) or base (liming). It acts as a reservoir of cations (Ca^{+2} , Mg^{+2} , K$^+$ and micronutrients) and anions (PO$_4^{-3}$ and SO$_4^{-2}$), favoring physical conditions such as aggregation and aggregate stability, aeration, water retention capacity and soil permeability, reducing susceptibility to erosion (SIQUEIRA E FRANCO, 1998).

Many conceptual models separate organic matter according to its stability and speed of decomposition by the action of soil microorganisms, which results in the emission of CO$_2$ and a change in the soil's chemical composition. Biological activity converts the leaf litter or straw into

7

stable humus, improving the aeration and physical aspects of the soil by incorporating the organic matter into deeper layers (KUTSCH et al., 2010).

In this way, the added organic matter not only influences soil respiration directly through its decomposition, but also creates ideal conditions for soil microorganisms and plants, improving the soil's physical conditions, determining its properties, and consequently other environmental variables correlated with soil CO_2 efflux.

Soil carbon saturation has been reported in different types of soil, with different textures and under different climates (STEWART, 2009). The process occurs mainly in the surface layers, due to the accumulation generated by leaves, branches and surface roots (NICOLOSO, 2009), and is represented by an asymptotic model (Figure 3), for the carbon stock, carbon addition relationship, rather than a linear model (SIX et al., 2002).

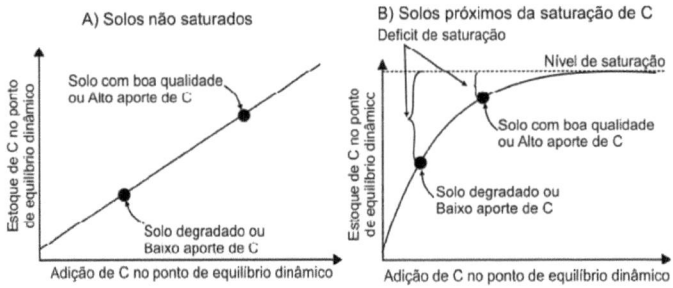

FIGURE 3: THEORETICAL MODEL REPRESENTING THE RESPONSES OF SOILS WITH DIFFERENT LEVELS OF DEGRADATION. SOURCE: BAYER ET AL. (2011).

Kinetic models that consider the accumulation of organic matter in the soil to be linear may overestimate the capacity of the soil to retain it, and disregard the saturation process (Figure 3) (NICOLOSO, 2009). Saturation occurs in carbon protection mechanisms (CHUNG et al., 2008). The linear model is efficient for representing carbon addition in degraded soils. In saturated soils, the asymptotic model adequately represents the accumulation of organic matter.

Degraded soils with low carbon contents have the greatest capacity and efficiency for storing carbon (Figure 3), as they are far from their saturation level. Carbon-13 tests have shown that the greater the deficit, the greater the capacity to stabilize the added carbon, and that the stabilization efficiency decreases with the increase in carbon in the soil (STEWART et al., 2008).

It can be seen that the addition of carbon is maximum in tropical forests and cultivated soils, where phytomass production is higher than in temperate forests and tropical savannas, which have climatic or nutritional limitations. The rate of decomposition (K) is greatly influenced by environmental factors such as temperature, humidity and aeration, varying considerably between ecosystems and being higher in cultivated soils or under tropical forests (SIQUEIRA E FRANCO, 1988).

The main physical changes that occur in the soil of cultivated areas compared to the soil of native forests are a decrease in macro porosity, total porosity and saturated hydraulic conductivity, as well as an increase in soil density (ZALAMENA, 2008). A high soil density limits the amount of oxygen available to microorganisms. In contrast, high porosity favors oxygenation of the soil, promoting microbial activity and consequently increasing emissions (FANG et al., 1998).

The ability to protect and stabilize carbon in the soil, in addition to the management practices adopted, depends on the intrinsic characteristics of the soil. Clay soils are more efficient at stabilizing and conserving soil carbon than sandy soils (GREGORICH et al., 1995; BOLINDER et al., 1999). A positive nitrogen balance is also essential for tropical and subtropical soils to be efficient at accumulating organic matter (URQUIAGA et al., 2010).

The carbon stock will depend on the type of vegetation present on the site, the quality and quantity of plant material that each species produces and deposits on the soil, and the climate will determine the speed of decomposition, consequently the CO_2 emissions from the topsoil into the atmosphere. Tropical and subtropical species are efficient producers of biomass.

Grass species, such as brachiaria, have an enormous capacity for carbon production, producing over 26 t ha^{-1} of dry matter, comparatively more than other crops. Millet, for example, produces 8 t ha^{-1} of dry matter (KLUTHCOUSKI AND AIDAR, 2003; KLUTHCOUSKI AND STONE, 2003). A native semi-deciduous forest in the state of Sao Paulo produces 12.2 tons of dry matter per hectare per year, including leaves and branches (HORA et al., 2008).

In seasonal semi-deciduous forests, the percentage of deciduous trees, i.e. those that lose all their leaves in winter and deposit organic matter in the soil, is between 20 and 50% of the total number of individuals (VELOSO et al., 1991). In the region of Limeira - SP, in a reforested area, leaf litter production was higher in winter (697 kg/ha) than in summer (407 kg/ha), showing a strong seasonal variation, which is a strong indication of the degree of growth and ecological balance of the new forest (MOREIRA E SILVA, 2004). CO_2 emissions are therefore expected to be one of the indicators of the environmental quality of forest systems.

The roots of plants are more efficient at accumulating carbon in the soil than the leaves, branches and other components of the aerial part. This explains why grass species are often as efficient as forests at accumulating carbon in the soil. In a comparative study, the roots converted 21% of their biomass production, while the aerial part converted only 12% (BOLINDER et al., 1999).

Roots, during their growth and after their senescence, contribute to the formation and stabilization

9

of soil aggregates, increasing carbon accumulation rates through the physical protection of organic matter (DENEF AND SIX, 2006), and the type of plant root system influences the formation and stabilization of macroaggregates (GALE et al., 2000).

Soils with clayey surface horizons are more efficient at stabilizing and conserving carbon in the soil when compared to sandy soils (GREGORICH et al., 1995; BOLINDER et al., 1999), 1999), showing lower rates of organic matter degradation, thus a Latossolo Bruno (with 620 g kg^{-1} of clay), evaluated in both conventional and no-till, had a decomposition rate of 1.4% and 1.2% for each type of planting, respectively (BAYER et al., 2006). Loose-textured Argisols had a decomposition rate of 3.14 % for conventional tillage and 1.82 % for no-till (LOVATO et al., 2004).

Organic matter in tropical clay soils is generally associated with iron oxides, due to the high chemical stability of the organomineral reaction, while soils with high clay contents show low degradation rates, even after the surface layers have been disturbed (OADES et al., 1989).

Microscopy has shown that carbon, when adhered to the colloidal fraction of clay, is protected from decomposition by microorganisms (RAZAFIMBELO et al., 2008). Consequently, the stabilization of organic matter in the soil depends on its texture and mineralogy, so silt and clay contents are reliable parameters for determining the stabilization capacity of organic matter in the soil (HASSINK et al., 1997).

Soil quality can be divided into dynamic and inherent. Attributes such as texture and mineralogy are innate to the soil and are determined by the length of time it has been exposed to the climate, the source material and the relief. These factors define soil quality. Anthropogenic activities alter the soil's physical, chemical and biological characteristics, defining its dynamic quality (PEIXOTO, 2008). It is not easy to select a set of properties that meet all the conditions for properly evaluating soil (LI and LINDSTROM, 2001).

2.3 Soil CO2 Emissions

The efflux of CO_2 from the soil began to be called "soil respiration" in the 1920s by Swedish researcher Henrik Lundegardh, who was responsible for the first measurements using a "static closed chamber" (KUTSCH et al., 2010). Soil respiration corresponds to the CO_2 produced by the respiration of roots, soil microorganisms and the aerobic decomposition of OM, and is a process that varies with vegetation and soil type (DAVIDSON et al., 2002), the result of physical, chemical and biological processes, influenced by soil moisture and temperature (EPRON et al., 2006; OHASHI AND GYOKUSEN, 2007) air temperature, air humidity, photosynthetically active radiation (LLOYD AND TAYLOR, 1994; DAVIDSON et al., 1998). Other factors that affect soil respiration are: bacterial activity (LLOYD AND TAYLOR, 1994), phosphorus content (DUAH-

YENTUMI et al., 1998), C/N ratio (ALLAIRE et al., 2012) and pH (FUENTES et al., 2006).

The carbon produced by root respiration is called "autotrophic" carbon, while that produced by litter decomposition is called "heterotrophic" (KUTSCH et al., 2010). "Autotrophic" respiration can be separated into plant root respiration, symbiotic mycorrhizal respiration and rhizosphere microbiota (KUTSCH et al., 2010). It is estimated that this "autotrophic" respiration is responsible for 40%-70% of the total CO_2 efflux from the soil to the atmosphere (HANSON et al., 2000; BOND-LAMBERTY et al., 2004; SUBKE et al., 2006).

Physical mechanisms also influence the efflux of carbon from the soil. Rommel (1922) observed that diffusion, due to the CO_2 gradient, is the driving force that takes the air mass from the soil layers to the atmosphere. Albertensen (1977) listed other factors and physical aspects that influence CO_2 efflux through the soil, such as: temperature, which induces differences in density and diffusivity between soil and atmospheric air; changes in barometric pressure; displacement of air in the soil due to water percolation (rain, irrigation); changes in the height of the water table; dissolution and transport of gases originating from liquid effluents; and pressure changes induced by wind speed (KUTSCH et al., 2010).

Information on the influence of humidity and temperature on the activity of soil biota, as well as pH and nutrient availability, has been known since the mid-nineteenth century (KUTSCH et al., 2010). Sotta (1998) cited five factors that can control the rate at which CO_2 is emitted from the soil into the atmosphere: its production rate in the soil, temperature gradients, the concentration at the soil-atmosphere interface, the soil's physical and chemical properties and fluctuations in atmospheric pressure.

Empirical relationships between CO_2 fluxes and environmental variables show that when there are no limiting factors, such as soil moisture, silt/sand/clay ratio, density and other physical properties of the soil, carbon emission increases exponentially with temperature (RAICH & SCHLESINGER, 1992). Under high temperature conditions, soil respiration is reduced by restricting microbial activity, and temperature also affects the speed of enzymatic reactions of the soil microbiota (KANG et al., 2003).

Among the physical factors that influence emissions, diffusion is the main one (VAL BAVEL, 1951, 1952). Some studies have shown the influence of wind speed on emissions, but there is a lack of in-depth and systematic research on the subject (KUTSCH et al., 2010). Thus, CO_2 exchange in soil-vegetation-atmosphere systems is directly and indirectly associated with meteorological events, which suggests that meteorological data alone could explain a significant part of the temporal variability in CO_2 emissions from soils (LA SCALA et al., 2003).

11

A number of studies and surveys have been carried out in recent years to characterize CO_2 efflux through the soil in the most diverse biomes on the globe. This effort seeks to understand the processes that influence the global carbon balance and, consequently, global warming.

Measurements of CO_2 emissions in the province of Shannxi, China, in an area located at an altitude of 1353 meters, with an annual rainfall of 504 mm and an average temperature of 10.1 °C, showed average annual values of 3.23 μmol CO_2 m s^{-2-1} , for a forest dominated by Liaodong oak (*Quercus liaotungensis*), 2.29 μmol CO_2 m s^{-2-1} for an oriental plane tree forest (*P. orientalis*), 2.35 μmol CO_2 m s^{-2-1} in an *acacia-bastard* plantation (*Rpseudoacacia*) and 2.03 μmol CO_2 m s^{-2-1} for a deforested area (SHI et al., 2014).

In temperate climate conditions, in an area in Slovakia, emission values varied during the seasons, ranging from 0.92 in winter to 15.20 μmol CO_2 m s^{-2-1} in summer for forest areas, and from 0.96 to 12.92 μmol CO_2 m s^{-2-1} in areas covered by grass (PRIWITZER, 2013). Other temperate forest ecosystems have also shown lower emission values during winter than during summer, 0.64 μmol CO_2 m s^{-2-1} in the Austrian winter (SCHINDLBACHER et al., 2007) and 0.67μmol CO_2 m s^{-2-1} in the cold season in Washington state, United States (MCDOWELL et al., 2000).

Still in a temperate climate, in Croatia, a study of correlations between meteorological variables and CO_2 emissions recorded a positive correlation with soil temperature ($r^2 = 0.42$) and air temperature ($r^2 = 0.45$) and a strong negative correlation with air humidity ($r^2 = -0.55$) (BILANDZIJA et al., 2014).

In Brazil, in native forests in the Amazon biome, they found average emission values of 6.4 μmol CO_2 m s^{-2-1} in the city of Manaus - AM, (SOTTA et al., 2004) and 6.1 μmol CO_2 m s^{-2-1} in the municipality of Paragominas - PA, (TRUMBORE et al., 2006). Some authors have found lower values for the northern region of the country, 3.2 μmol CO_2 m s^{-2-1} in Manaus (CHAMBERS et al., 2004) and 4.25 μmol CO_2 m s^{-2-1} in Juruena, state of Mato Grosso (NUNES, 2003).

In the Amazon rainforest, significant relationships ($p<0.05$) were found between CO_2 emissions and soil moisture, in Sinop-MT during the dry season ($R^2 = 0.76$) and the rainy season ($R^2 = 0.78$). In Caxiuana, a significant relationship was also found between the variables during the dry season ($R^2 = 0.82$) and the rainy season ($R^2 = 0.82$). The same occurred in Manaus-PA with significant values for the dry season ($R^2 = 0.68$) and the rainy season ($R^2 = 0.60$) (DIAS, 2006).

The correlation between soil moisture and CO_2 emissions by the soil has already been demonstrated by various authors. For Dias (2006), in general, carbon flows into the atmosphere are greater during the rainy season than during the dry season, with soil moisture and temperature being the main

12

factors conditioning the production of this gas by the soil.

In tropical forests, several authors have found a significant positive linear correlation between soil respiration and soil temperature (EPRON et al., 2006; DIAS, 2006). On the other hand, in an area cultivated with sugar cane in the interior of São Paulo, emissions showed no significant correlation with soil temperature (BICALHO et al, 2014), which can be explained by the low variability of the variable during the collection period (DIAS, 2006).

In the state of Sao Paulo, unfortunately there are no studies on the Atlantic Forest, the existing records were obtained in areas of sugar cane cultivation, and the average values measured are: 1.5 μmol CO_2 m s^{-2-1} after mechanized harvesting (BISCALHO et al., 2014). Brito et al. (2010) point out that CO_2 emissions in sugarcane cultivation can vary depending on the topography and types of management employed, as previously observed by Panosso et al. (2009), who measured emissions of 2.16 μmol CO_2 m s^{-2-1} in areas where harvesting was mechanized and 5.29 μmol CO_2 m s^{-2-1} for areas with manual harvesting, preceded by burning the sugarcane.

In an area planted with sugar cane, he found daily averages of between 1.26 and 1.77 μmol CO_2 m s^{-2-1} during the month of July in the city of Guariba in the interior of Sao Paulo. The coefficients of variation ranged from 40% to 90%. And a significant positive linear correlation ($p<0.05$) with macroporosity ($r^2 =0.21$) and negative with microporosity ($r^2 =-0.18$) and soil density ($r^2 =-0.32$) (BICALHO et al., 2014).

Significant linear correlations between CO_2 emissions and soil attributes such as macroporosity, microporosity and density have been cited by several authors (EPRON et al., 2006; PANOSSO et al., 2011; TEIXEIRA et al., 2013; BICALHO et al., 2014), suggesting the importance of these attributes as regulators of microbial activity and, consequently, soil CO_2 emissions.

The thermal properties of soils have also been correlated with emissions, in a monitoring of CO_2 emissions in a pasture in the state of Missouri - USA, a significant correlation was found ($r^2 =0.62$, $p<0.0001$) between soil respiration and thermal conductivity (NKONGOLO et al., 2010).

13

3. CHARACTERIZATION OF THE STUDY AREA

It is estimated that the state of Sao Paulo originally had 81.8% of its area covered by forests (20,450,000 ha). Studies on the evolution of forest cover show that, in 1990, only 1,731,472 ha remained, or 4.16% of the state's territory. Of this total, 45.77% (792,448.57 ha) are part of Conservation Units (UCs) under the responsibility of the Department of the Environment (SÂO PAULO, 1998).

The study area, the Edmundo Navarro de Andrade State Forest, is located in the municipality of Rio Claro and is a sustainable use PA, created by State Decree 46.819, in accordance with Law 9.985/00, which established the National PA System. The municipality, located 173 km northwest of the capital of the state of São Paulo, has two districts, Assistência and Ajapi (Figure 4), a total area of 499.9 km^2 , and is part of the Piracicaba urban agglomeration and the Corumbatai River basin, which can be accessed via the Anhanguera/Bandeirantes system and the Washington Luiz Highway (SP 310).

The forest, located on the eastern edge of the urban area of the city of Rio Claro, was created in 1909 and covers an area of 2,230.5 hectares. It has the largest variety of eucalyptus species concentrated in a single area in Brazil, making it a benchmark in forest cultivation, research and production, and internationally known as the "cradle of eucalyptus" (IF, 2005).

It originally belonged to CPEF-Companhia. Paulista de Estradas de Ferro, and was transferred to FEPASA-Ferrovia Paulista S.A. during the 1970s, when the railroads were nationalized. Since 1998, it has been administered by SMASP-Secretaria de Meio Ambiente do Estado de São Paulo, with FF-Fundaçao Florestal being responsible for the management of the unit (IF, 2005).

It is estimated that there are still more than sixty species of eucalyptus in FEENA, as well as spontaneous and induced hybrid species. This entire area constitutes an important genetic bank, with strategic value in the event of the introduction of a new pest or disease unknown to Brazilian forestry. Edmundo Navarro de Andrade, the creator of FEENA, was much criticized by nationalists, who did not agree that the introduction of eucalyptus would lead to superior wood quality and faster growth than local species (IF, 2005).

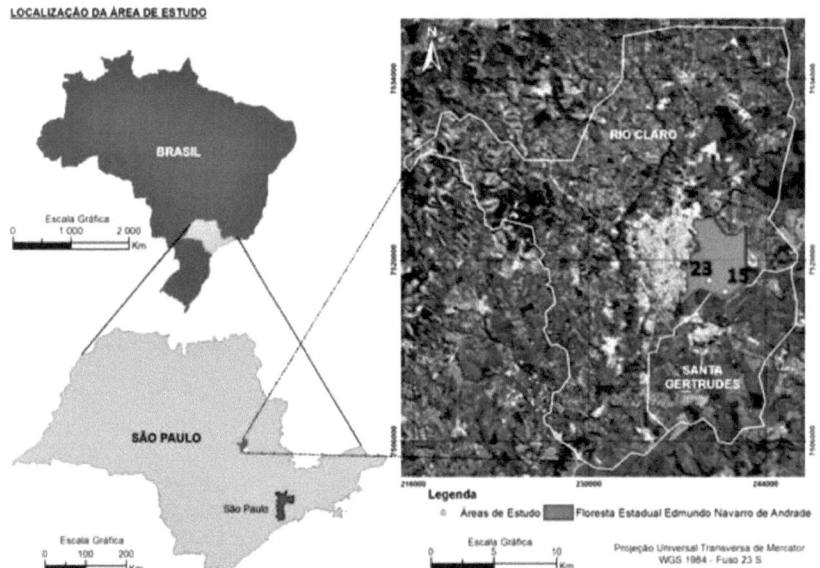

FIGURE 4 - LOCATION OF FEENA AND THE PLOTS (15 AND 23) WHERE THE CO2 EMISSIONS SURVEYS WERE CARRIED OUT.

In order to make the conservation of the eucalyptus genetic base, native vegetation and public use compatible, the UC was spatially organized into zones and woodlots, according to the different uses and degrees of protection required (IF, 2005). By cross-referencing the basic surveys with data from fieldwork and other available information, FEENA's woodlands were classified into zones called: Historic-Cultural, Recovery, Forest Management, Conflict, Public Use, Special Use, Conservation. Each one has different rules for use, determining the different functions, be they social, administrative, ecological, management or protection, for each of FEENA's spaces.

The Historic-Cultural Zone contains historical, scientific, cultural and archaeological samples that must be preserved and interpreted for the public. Its aim is to protect historical and archaeological sites in harmony with the environment, promote scientific research, environmental education and interpretation. This area includes the old coppices, which mark the beginning of the plantations (IF, 2005).

The largest area of the unit is the Forest Management Zone, which comprises native or planted forests with economic potential for multiple and sustainable resource management. The aim is to generate technology and forest management models, with research, environmental education and interpretation activities. The purpose of the Public Use Zone is intensive recreation, leisure and environmental education in harmony with the environment (IF, 2005).

Degraded areas are called Recovery Zones, and once they have been recovered they will be

15

incorporated back into one of the other permanent zones. Its aim is to stop the degradation of resources, and it may also include research, environmental education and interpretation activities (IF, 2005). Included in the Special Use Zone are the areas needed for administration, such as the headquarters, staff housing in the colonies and the Military Police kennel (IF, 2005).

Areas occupied by public utility developments are called Use Zones, places with gas pipelines, oil pipelines, transmission lines, antennas, water catchments, dams, roads, optical cables and others (IF, 2005).

The areas subject to this study, plots 23 and 15 (Figure 4), are located in the Historic-Cultural and Forest Management Zones, respectively. Plot 23 is correctly included in the Historical-Cultural Zone as it is a historical, scientific and cultural sample of one of the first plots planted with native species in Brazil. Talhao 15, on the other hand, is located in the Forest Management Zone and its current use, recovery and environmental conservation is in line with what is established in the plan, which is commercial exploitation and the multiple and sustainable use of forest resources.

3.1 Characterization of the physical environment at FEENA

FEENA is part of the Corumbatai river basin, whose main tributaries are the Passa Cinco, Cabeça and Ribeirao Claro rivers. The headwaters are located on the escarpments of the Serra dos Padres basaltic ridgeline, and its waters flow into the Piracicaba River. The UC's surface water bodies are made up of small streams, such as the Ibitinga and Santo Antônio streams, and the main stream, the Ribeirao Claro, is used to collect water for the municipality (IF, 2005).

The area in which the Ribeirao Claro Basin is located is characterized by the presence of tabular interfluves, stepped terraces and plains, at altitudes of between 550 and 650 metres (PENTEADO, 1968). The slightly dissected aspect of the Basin is due to the streams that carve its valleys, generating gentle slopes that delimit the subtabular interfluves that dominate the region (PENTEADO, 1981).

The Ribeirao Claro crosses the UC in a north-south direction, establishing in some stretches the boundary between FEENA and the urban area of Rio Claro. This river flows through an open valley with a flat bottom, where there are well-developed river plains and abandoned meanders, forming alluvial deposits of sand and clay (IF, 2005).

The Forest is located in the relief compartment of the state called the Paulista Peripheral Depression, a geomorphological unit whose origin is linked to the establishment of a zone of structural weakness in the contact between sedimentary lithologies linked to the Paranà Sedimentary Basin and Precambrian lithologies associated with the Atlantic Plateau (IF, 2005).

16

Geologically, the two areas selected for the field surveys are based on basal intrusive rocks associated with the Paranà Magmatic Province (PMP), considered to be one of the largest volcanic manifestations of a basal nature in the continental area of the Earth, involving, in Brazilian territory, the states of Rio Grande do Sul, Paranà, Santa Catarina, São Paulo, southwest Minas Gerais and southeast Mato Grosso do Sul. Basalts occur in the form of effusions and intrusive rocks (sills and dykes) (MACHADO et al., 2007).

The soils in the study areas are called Red Argisols due to the color given by the high levels and nature of the iron oxides present in the original material. Their natural fertility depends on the source material. As it is classified as Eutrophic, it is a soil of good fertility. The clay content in the subsurface horizon (red in color) is much higher than in the surface horizon, and this increase in clay is easily perceived when the texture is examined in the field (EMBRAPA, 2006).

As they are classified as typical, in the fourth level of trophic classification, the soils in the study areas do not have any restrictive characteristics which could limit agricultural activities, such as abrupt soils, in which the textural difference between the surface horizons makes the soil susceptible to erosion, or saprolitic soils which restrict root penetration into the surface (EMBRAPA, 2006). The native vegetation on the site is the Semideciduous Seasonal Forest, which covers well-drained eutrophic basaltic soils in the interior of the state of São Paulo (RODRIGUES, 1999).

The study area is part of the Atlantic Forest biome and is dominated by Seasonal Forests, also known as Mesophytic Forests. Unlike the Ombrophilous Forests (moist and evergreen), the Seasonal Forests are governed by a marked climatic seasonality, with the percentage of deciduous trees reaching 50%. In Rio Claro, the seasonal forests are often interspersed with Cerrado formations, a domain which in this region is determined by sandy soils with low water retention capacity (IF, 2005).

The climate in the FEENA area is classified as Koppen's Cwa: *mesothermal* (with an average temperature of the coldest month between -3 °C and 18 °C) and *high altitude tropical* (with a dry winter and an average temperature of the hottest month above 22 °C). The average annual temperature is 20.6 °C (Figure 8) and a distinction can be made between the hottest period (September to April), with an average of over 22 °C between December and March, reaching 23 °C in February; and the least hot period (May to August), with temperatures below 19 °C, with June and July being the coldest months (17.1 °C) (IF, 2005).

Annual rainfall is 1,534 mm, with two distinct seasons: a rainy period from October to March, when rainfall reaches 1,188 mm (77% of the total); and a drier period from April to September, with an average rainfall of 346 mm (23% of the total). The rainiest months are also distinguished

17

(December, January and February): 248, 252 and 210 mm respectively; and the least rainy months (June, July and August), 48, 34 and 34 mm respectively (IF, 2005) (Figure 5).

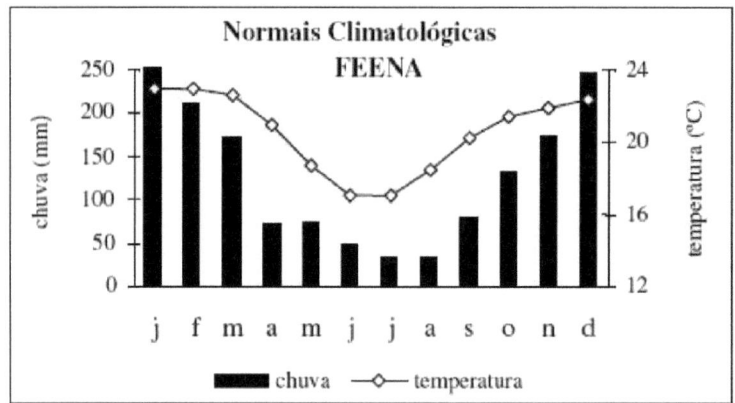

FIGURE 5. CLIMATOLOGICAL NORMALS FOR RAINFALL AND PRECIPITATION MEASUREMENTS FROM 1954 to 1997.

SOURCE: IF (2005).

the rainfall regime is influenced by the Atlantic Tropical and Continental Equatorial masses, which bring moisture to the continent. High temperatures cause warm, humid air to rise, leading to rainfall. The relief of the cuestas causes orographic rainfall, also contributing to the high rainfall. In winter, low temperatures are influenced by the Atlantic Polar Mass (MONTEIRO, 1967).

According to the climatological water balance (THORNTHWAITE AND MATHER, 1955) (Figure 6), the annual water deficiency is only 7 mm, concentrated in July and August. The annual water surplus is 572 mm, concentrated between October and March. In the other months there is no or almost no surplus (IF, 2005).

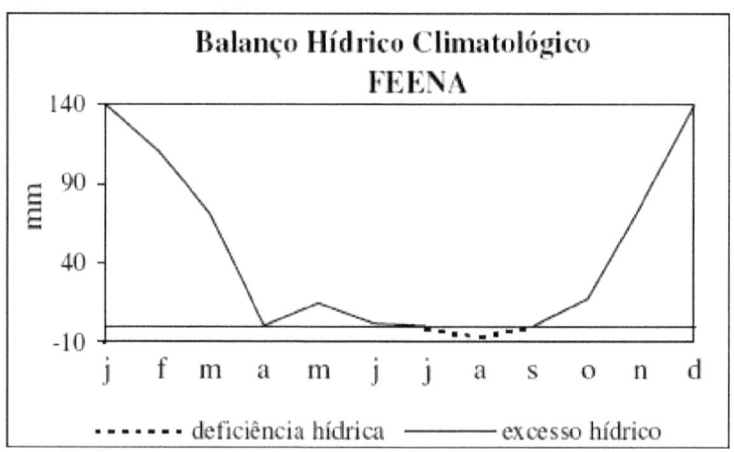

Figure 6 - Graphical representation of the water balance and climatology from 1954 to 1997.

Source: IF (2008).

4. MATERIALS AND METHODS

4.1 Area selection and experimental design

In order to evaluate and characterize the differences between soil carbon emissions in areas that have already been restored and in the process of being restored within the Atlantic Forest morphoclimatic domain, we selected a newly planted area, plot 15, planted in 2014, and an almost century-old area.

plot 23 (Figure 7), was planted by Navarro de Andrade in 1916 with the aim of comparing the growth of these trees with eucalyptus, and demonstrating that the Australian species grew faster and had superior wood quality for the production of charcoal, firewood and sleepers. Seedlings of 70 species from 25 different families, many of commercial interest, were planted in this plot, spaced 2m by 2m over an area of 1.1 hectares, proving the initial idea of its creator, that is, in comparing the growth of the natives with eucalyptus, it was observed that the exotic species was the most suitable for large-scale planting by the Companhia Paulista de Estradas de Ferro (IF, 2005).

FIGURE 7. PARTIAL VIEW OF CUT 23 of THE ACCESS ROAD.

As almost all of FEENA has been reforested with eucalyptus or other exotic species, areas with native vegetation exist as a result of restricted or non-existent forest management processes (historical collections and plots of interest for genetic improvement), or the absence of occupation of previously forested areas (abandoned plots). In these cases, *native vegetation* can be found, either through the formation of an understorey in the older plots, or through regeneration, infestation or seed rain from neighboring forested areas (iF, 2005).

Recently, some of these abandoned plots at FEENA were included in a commitment to environmental recovery and restoration planting was carried out. In 2014, plot 15 (Figure 8) was replanted with more than 80 different species, in line with SMA Resolution 8 of 31-01-2008, which set out the guidelines for heterogeneous reforestation of degraded areas, and the planting was monitored by cETEsB, as it involved environmental compensation.

FIGURE 8. DETAIL OF THE EXPERIMENTAL PLOT INSTALLED ON PLOT 15.

Since the beginning of the century, the land has been occupied by eucalyptus plantations on plot 15, which is part of the Forest Management Zone. After the last cut, approximately 10 years ago, the site was abandoned and has since been occupied by grass species such as colony grass.

Recently included in an environmental recovery agreement and replanted, its destination is in conflict with the Management Plan. It is closer to an area cultivated with sugar cane than a forest, due to the straw from the recently desiccated grasses on the ground, and the history of machinery traffic during the harvesting of the various eucalyptus cycles.

In order to assess CO_2 emissions from the soil in these two areas, 900 m sample plots were installed[2] . Within these plots, 17 collection points were installed, distributed as shown in figure 9. The distance between the points was set at 10 meters (twelve points), while in the central portion the distance between the points was 5 meters.

21

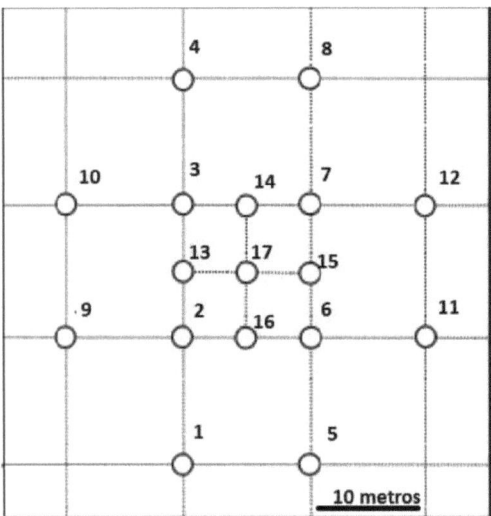

FIGURE 9. DISTRIBUTION OF SAMPLING POINTS.

In order to fix the equipment with which the measurements were taken, a PVC ring 10 to 15 cm high was fixed to the ground at each of the points (Figure 10), 48 hours before the measurements were taken, remaining installed throughout the collection period, in order to minimize changes in the structure of the litter and the soil surface.

The measurements were taken between September 2014 and May 2015, between 08:00 and 17:00 hours, in order to take advantage of the time of day when there is the most sunlight, increasing the safety of the work and the people involved. Five flow measurements were taken at each point, along with the collection of data on environmental variables: air pressure, temperature and humidity. The physical properties of the soil were measured once at each point, at the same time as the flow data was collected. The variables measured were: humidity, temperature and the coefficient of thermal resistivity of the soil. The soil for determining the C/N ratio was collected in September 2014.

FIGURE 10. COLLECTION RING INSTALLED IN PLOT 15, CAMERA BEING ATTACHED TO THE RING IN PLOT 23.

4.2 CO_2 flux measurements

CO_2 flow measurements were carried out using equipment developed by Moreno (2012) at UNESP-Rio Claro, consisting of an infrared gas analyzer (iRGA), model Li-840, brand Li-Cor, coupled to a dynamic chamber using a circulation pump (Figures 11 and 12).

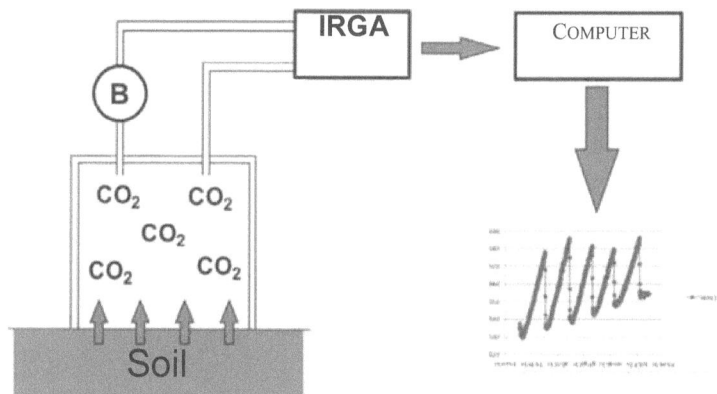

FIGURE 11. DYNAMIC CHAMBER TOGETHER WITH AN INFRARED GAS ANALYZER (IRGA) AND a pump (B) to force the gas to circulate through the IRGA. SOURCE: MODIFIED FROM MORENO (2012).

Advantageously, this system represents an alternative to the various commercial systems available for this purpose. Among these advantages are: the low total cost and maintenance of the system, the possibility of automatic or remote control via the internet, the possibility of changing the detector for measurements of other gases and the simultaneous measurement of other parameters such as humidity, temperature, pressure and air velocity at the sampling site (MORENO, 2012).

The flux due to soil respiration is calculated as the rate of change of CO_2 concentration within the chamber volume per unit time, according to equation (5) below:

$$Rs=(Cn-Cn1)/\Delta t*(V/A)*(P/RT), \qquad (5)$$

Where: Rs=Referent CO_2 flux (μmol m^{-2} s^{-1}), Cn=CO2 concentration (ppm), P=Air pressure (Pa), T=Air temperature (K), R=Specific gas constant (8.314 J mol^{-1} K^{-1}), V=Chamber volume (m^3), A=Chamber horizontal coverage area (m $)^2$

Infrared spectroscopy, the analytical method used by the equipment to determine CO_2 concentrations, uses the absorption of radiation to measure the concentration of chemical compounds and is usually used to determine the concentrations of compounds made up of hydrogen, carbon or oxygen and nitrogen (MORENO, 2012).

The infrared gas analyzer (IRGAS) (Figure 13) consists of an infrared emitter, a measuring cell (called an optical path), an optical filter and a detector. The infrared signal from the source passes through the measuring cell where the gas sample to be analyzed is located. Before hitting the sample, the light passes through a monochromator (which can be a prism, a dispersion network or a filter), which transforms the polychromatic light into monochromatic light (ROMANO, 2006, apud

24

MORENO, 2012).

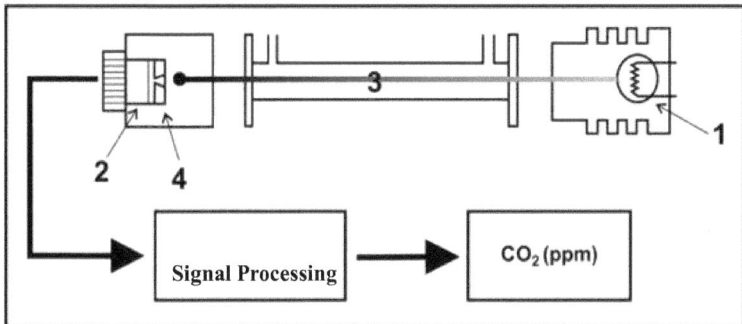

FIGURE 13. BASIC COMPOSITION OF THE IRGA.1. INFRARED SOURCE, 2. DOUBLE DETECTOR, 3.

SAMPLE CELL (OPTICAL PATH), 4. FILTER. SOURCE: MORENO (2012).

As the air sample passes through the analyzer, in this case the Li-840, it is irradiated by a light beam of known intensity (P0). The irradiated photons come into contact with the molecules in the sample, and when these have vibrational energy incompatible with the energy of the photons, no energy is absorbed and all the photons pass through the sample. In this case, the irradiated beam leaving the sample will have the same intensity as the incident beam P0 = P. Similarly, if the energy of the photons in the irradiated light is compatible with the vibrational energy of the molecules, they will absorb the photons, increasing their vibrational movement, and consequently the intensity of the incident beam will be reduced. The intensity of the photon beam leaving the sample will be lower than the initial incident intensity (P0 > P) (HARRIS, 1999 apud MoRENo, 2012).

Before taking the field readings, the equipment was calibrated at the Physics Department of UNESP Rio Claro. Two gas mixtures with known concentrations were used and the calibration polynomials available in the device's software were applied. The calibrations were made with a mixture containing only pure nitrogen, therefore with 0 ppm CO_2 (0% CO_2), and another with a concentration of 335 ppm CO2 (0.035% CO_2).

The acquisition software used to record the data collected by the Li-840 is also used to calibrate the analyzer, so it is possible to calibrate and record the level with no CO_2 concentration (*Zero CO_2*), and the *Span CO_2* level where the known concentration is recorded. In this way, two points with a known concentration are required for calibration.

After calibrating the equipment, the data was collected at FEENA. Each data acquisition curve shown in the graph (Figure 14) provides information on the CO_2 concentration in relation to time, using this information and equation (5) to calculate a CO2 emission measurement. Up to five measurements were taken at each point, and each of the curves shown in figure 14 was obtained by closing the equipment's chamber and accumulating CO2 inside it. When the chamber is opened, every 2 minutes on average, there is an abrupt decrease in the concentrations inside the chamber.

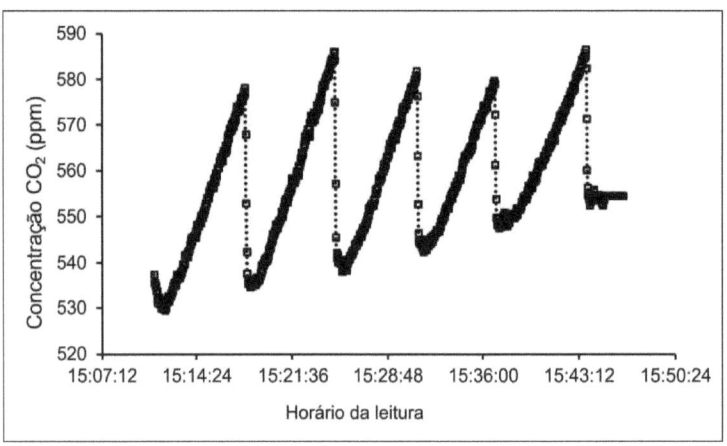

FIGURE 14: CO2 ACCUMULATION MEASUREMENTS AT POINT 17 OF PLOT 15.

4.3 Soil Moisture

To measure soil moisture in the field, we used a device called the "*Speed moiusture tester*" (Figure 15), which, according to Garzella (2011), gives satisfactory results in determining moisture in various types of soil. This equipment was initially used for quick determinations of materials of various origins, such as seeds, fiber and coal, based on the reaction of water with carbide.

FIGURE 15. SPEED-TYPE SOIL MOISTURE METER.

The chemical principle behind the *Speed* meter derives from the process of forming and quantifying acetylene by reacting water with calcium carbide, also known as the calcium carbide method. The principle of measurement consists of mixing calcium carbide with the material to be analyzed inside a cylinder, and from the reaction with the water present in the soil, acetylene gas is formed. In this process, the water contained in the material to be analyzed promotes the hydrolysis of the carbide, causing two hydrogen atoms to replace the calcium in its structure, giving rise to acetylene, according to the chemical reaction below (GARZELLA, 2011):

$$2\ H_2O + CaC_2 \rightarrow Ca(OH)_2 + C_2H_2\ (\uparrow) + \text{energy (6)}$$

This establishes a stoichiometric relationship between the amount of water used as a reactant and the amount of acetylene obtained as a product. Based on this relationship, where each mole of acetylene corresponds to two moles of water, it is possible to determine the water content of a sample by quantifying the acetylene formed. As it is a gas at room temperature, its quantification is carried out by measuring the pressure exerted by it on the inside of the cylinder, using a manometer (GARZELLA, 2011).

Using tables, the pressure value obtained is converted into the percentage of water contained in the sample. When taking readings, it is often difficult to obtain the correct moisture content due to problems in reading the pressure, or due to the frequent lack of correspondence in the pressure and moisture conversion table (GARZELLA, 2011).

Initially, it was necessary to calibrate the device's readings, which was carried out in April 2014, in order to optimize the handling of the equipment and obtain greater accuracy in determining soil moisture, allowing moisture to be correctly correlated with the other parameters measured in the project.

To carry out the calibration, one kilo of soil was collected from the 0-10 cm layer of plot 23, broken up and placed on a tray, exposed to the air so that it lost its natural humidity. Five 150 gram aliquots were then separated and placed in plastic bags, to which different amounts of water were added in order to obtain different humidities. These aliquots were stored in Styrofoam boxes for 3 days to homogenize them.

After all the soil aliquots had been homogenized, three samples of each treatment were placed in a crucible, previously weighed, and the wet weight was measured, the crucibles were placed in an oven at 100 C for 24 hours and then they were weighed again, from the difference between the wet weight and the dry weight the gravimetric humidity of the soil was calculated.

4.4 Soil temperature and thermal conductivity

Soil temperature and thermal conductivity were measured using a KD 2Pro data acquisition system (*Decagon*, USA), coupled to a KS-1 probe (a needle containing a heater and a thermocouple), with an accuracy of ± 5% for thermal conductivity values between 0.2 and 2.0 Wm K^{-1-1} and ± 1% for values between 0.02 and 0.2 Wm K^{-1-1}. The temperature and thermal conductivity data were collected 5 centimeters from the collection point by inserting the KS-1 probe into the soil during the acquisition of CO_2 efflux data (Figure 16). Although the KS-1 probe is not suitable for use in wet soil, it was used because it was the equipment available for measuring this parameter.

Figura 16.   KD2 - PRO EQUIPMENT COUPLED TO THE ÎHÎ SENSOR COLLECTING DATA IN PLOT 15.

4.5 Climatic parameters

The climatic parameters, temperature, air humidity and atmospheric pressure, were measured in the field with an ANOVA weather station, part of the DRIA-0511 model, placed on the ground next to the chamber, with continuous recording of these parameters during the CO_2 flow measurements. The measurements were recorded on a spreadsheet in the field and then correlated with CO_2 emissions.

4.6 Determination of soil carbon and nitrogen content

Soil samples were collected from plots 15 (17 points) and 23 (10 points) to determine the amount of carbon and nitrogen. The material was removed with a penknife, discarding the leaf litter, and the surface layer (0-5 cm) was collected. The material was then dried in an oven at 40°c, broken up manually with a wooden roller and passed through a fine mesh sieve (2 mm) to obtain the fine air-dried soil (TFsA).

The TFSA samples were macerated and passed through a ≤ 100 mesh sieve. Five grams of soil from each point were then separated, bagged and identified for analysis. Nitrogen was obtained using the Kjeldahl method (1883), and carbon was obtained using the Yeomans and Bremner method (1988).

4.7 Data processing

Multiple regression analysis was used to assess the correlation between the parameters measured in the field (independent variables) and CO_2 emissions (dependent variable). This is a multivariate statistical technique widely used in environmental studies to assess the predictive power of independent variables on dependent variables (HAIR JR. et al, 2005).

The generic multiple regression model is given by the expression below, when applied to a sample of size n (HAIR JR. et al, 2005):

$$Y_i = \beta_0 + \beta_1 X_{1i} + \beta_2 X_{1i} \ldots + \beta p X pi + \varepsilon_i, \quad i=1,2,..,n \quad (7)$$

Where,

Y_i = dependent or explained variable i=1, *2...n.*

β_0 = intercept or independent variable term

β_i = slope of Y with respect to variable X_i, keeping $x_2, x_3, \ldots x_p$ constant

βp = slope of Y with respect to variable X_p, keeping $x_i, x_2, \ldots x_{p-1}$ constant

ε_i = random error in Y, for observation i, i=1,2,......................,n.

The condition for multiple regression is that $\varepsilon_i \sim N$ $(0, \sigma^2)$, i.e. the errors must have a Gaussian distribution, be independent with zero mean and constant variance.

There are some statistical assumptions that cannot be violated when models are developed using multiple linear regression, and which are necessary for proper estimation. The modeling must meet at least the following assumptions: linearity, homoscedasticity and heteroscedasticity, independence of residuals, normality, *outliers,* collinearity and multicollinearity (HAIR Jr, et al., 2005).

To investigate the existence of any violation of the statistical assumptions of multiple linear regression, the simplest and most usual way is through the analysis of a residual graph (HAIR Jr. et al., 2009). Environmental data, such as that collected in this project, often present censored, missing and/or discrepant values (*outliers*), as well as not presenting a Normal or Log-normal distribution,

29

and the relationship between the measured and estimated values for the dependent variable can present large errors, known as heteroscedasticity, which can affect the prediction of the dependent variable (HAIR Jr. et al., 2005). When some of the statistical assumptions are violated, corrective action must be taken, in which case robust statistical methods may be the most appropriate to correct the violations of the general relationship (SABINO, et al., 2014).

Some minimum care must be taken with regard to the number of independent variables and the number of samples of the general relationship, the addition of a variable always increases the value of the relationship coefficient, when the sample number is small, this effect is called overfitting, this impact is minimized whenever the sample has a minimum of 10 to 15 observations per independent variable (HAIR Jr., et al., 2009).

5. RESULTS

The results obtained over the course of the project will be presented in this chapter, including laboratory activities (equipment calibration), field activities (CO_2 emission surveys) and statistical treatment of field data.

5.1 Calibration of the humidity meter

The humidities calculated for the soil samples prepared using the gravimetric method and the *speed* method are shown in Table 1 below, as well as the graph showing the correlation between these determinations (Figure 17).

TABLE 1: MOISTURES MEASURED IN THE SAMPLES PREPARED IN THE "SPEED" APPARATUS AND GRAVIMETRIC OF THE SAMPLES.

Sample	Speed" humidity	Gravimetric humidity
1	4,00	4,81
3	7,50	8,22
2	11,50	13,73
4	15,80	20,91
5	19,80	31,90

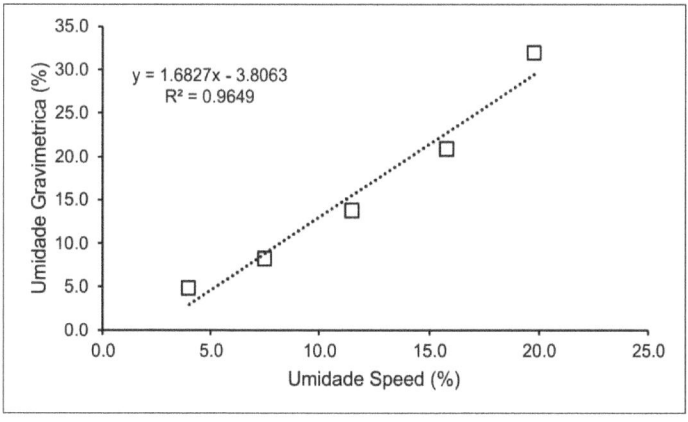

Figura 17. CORRELATION BETWEEN HUMIDITY MEASURED IN AN OVEN AND BY THE "*SPEED*" METER.

The correlation equation (equation 8) obtained made it possible to correct the moisture measurements made in the field, which were used in this study.

31

Y=1.68X - 3.806 (8)

where,

Y= *Speed* humidity and X = Gravity humidity

5.2 co2 **Emission Rates and Field Parameters**

The data relating to the co2 flow measurements taken are shown in tables 2 and 3, which contain the co2 emissions recorded, the date and time of recording, air temperature, air humidity, atmospheric pressure, soil moisture, soil temperature, thermal conductivity and the C/N ratio. Table 4 shows the basic statistics of the parameters evaluated for plots 15 and 23. A total of one hundred and twenty measurements of Co2 emissions from the soil were carried out, seventy-one on plot 15 and forty-nine on plot 23.

For the statistical treatment given to the set of data collected, it would be necessary to have the same number of data for both areas, but this was not possible. As a way of getting around this problem of sample distribution, only 49 samples were selected in plot 15, using the following criterion: from the average Co2 emissions at each point, only those with a smaller deviation from the average value were selected.

TABLE 2: CO EMISSIONS2 , AVERAGE ATMOSPHERIC PARAMETERS AND PHYSICO-CHEMICAL PARAMETERS OF THE SOIL IN PLOT 23.

Point	Date	Schedule	Emission (μmol CO2 m^ s^21 ·)	Umi. Air (°0)	Air temp. Air (°C)	P Atm (IrPa)	Humid. Soil (°0)	Soil Temp. Soil (°C)	Thermal Condition (Wm-1 K')1	C/N
1	7/10/2014	8:25	1.08	54	23.7	940.2	26.0	18.31	0.63	10.14
13*	10/7/2014	9:26	1.99	46	24.8	940.6	40.9	19.76	1.06	10.63
13	10/7/2014	9:33	2.29	41	24.8	940.8	40.9	19.76	1.06	10.63
13	10/7/2014	9:39	2.23	41	27.3	940.8	40.9	19.76	1.06	10.63
13	10/7/2014	9:45	2.01	29	33.3	940.8	40.9	19.76	1.06	10.63
13*	10/7/2014	9:53	2.59	33	33.3	940.8	40.9	19.76	1.06	10.63
3*	10/7/2014	10:11	1.99	30	30	940.6	37.7	19.41	0.97	10.27
3	10/7/2014	10:18	2.10	33	30.2	940.8	37.7	19.41	0.97	10.27
3	10/7/2014	10:25	2.17	35	28.5	940.5	37.7	19.41	0.97	10.27
3*	10/7/2014	10:32	1.92	35	28.5	940.5	37.7	19.41	0.97	10.27
3	10/7/2014	10:38	2.28	35	29.7	940.5	37.7	19.41	0.97	10.27
4*	10/7/2014	10:53	1.58	24	32.2	940.3	22.8	21.82	1.07	11.94
4	10/7/2014	10:59	1.79	24	35.8	940.1	22.8	21.82	1.07	11.94
4	10/7/2014	11:07	1.84	22	37.2	939.7	22.8	21.82	1.07	11.94
4	10/7/2014	11:14	1.93	24	35.5	939.7	22.8	21.82	1.07	11.94
4*	10/7/2014	11:21	1.98	29	31.3	939.6	22.8	21.82	1.07	11.94
5*	10/8/2014	14:32	1.39	32	37.8	940.6	21.6	33.02	0.41	12.63
5	10/8/2014	14:04	2.01	15	43.4	940.8	21.6	33.02	0.41	12.63

Point	Date	Schedule	Emission (μmol CO2 nr² s^1 ·)	Umi. Air (°0)	Air temp. Air (°C)	P Atm (IiPa)	Humid. Soil (°0)	Soil Temp. Soil (°C)	Cond. Tèrni (Wnr¹ K')¹	C/N
5	10/8/2014	14:48	2.06	12	46.2	940.5	21.6	33.02	0.41	12.63
5	10/8/2014	14:56	1.96	15	48.3	940.5	21.6	33.02	0.41	12.63
5*	10/8/2014	15:03	2.04	13	50.2	940.5	21.6	33.02	0.41	12.63
7	10/8/2014	16:04	1.61	13	36.3	934.6	31.9	23.45	0.92	8.57
7	10/8/2014	16:13	1.53	13	36.6	933.8	31.9	23.45	0.92	8.57
7	10/8/2014	16:02	1.33	13	36.6	934	31.9	23.45	0.92	8.57
8*	10/8/2014	16:45	E53	19	33.8	934	40.9	23.49	0.96	10.01
8*	10/8/2014	16:51	E87	21	32	934.2	40.9	23.49	0.96	10.01
8	10/8/2014	16:57	1.71	21	32	934.2	40.9	23.49	0.96	10.01
8	10/8/2014	17:04	1.71	23	31.1	934.2	40.9	23.49	0.96	10.01
8	10/8/2014	17:11	1.77	26	30.8	934.2	40.9	23.49	0.96	10.01
14*	10/9/2014	13:58	0.80	15	48.5	932.4	44.2	25.735	0.96	10.63
14*	10/9/2014	14:05	1.03	12	46.3	932.4	45.2	25.735	0.96	10.63
14	10/9/2014	14:12	0.96	13	45.3	932.4	46.2	25.735	0.96	10.63
14	10/9/2014	14:18	0.89	13	45.4	932.4	47.2	25.735	0.96	10.63
14	10/9/2014	14:25	1.01	12	46.9	932.2	48.2	25.735	0.96	10.63
15*	10/9/2014	14:38	1.50	13	45.9	932.1	27.5	35.295	0.86	13.63
15*	10/9/2014	14:44	1.69	12	46.5	931.9	28.5	35.295	0.86	13.63
15	10/9/2014	14:51	1.61	11	47.5	931.5	29.5	35.295	0.86	13.63
15	10/9/2014	14:57	1.68	12	46.9	931.9	30.5	35.295	0.86	13.63
15	10/9/2014	15:06	1.57	12	46.2	931.9	31.5	35.295	0.86	13.63
17	10/9/2014	15:13	1.15	14	44.3	931.9	33.2	29.14	0.73	10.63
17*	10/9/2014	15:02	1.52	12	46.2	931.9	34.2	29.14	0.73	10.63
17	10/9/2014	15:27	1.06	11	47.4	931.9	35.2	29.14	0.73	10.63
17	10/9/2014	15:32	0.98	15	48	931.9	36.2	29.14	0.73	10.63
17	10/9/2014	15:04	0.82	13	50	931.9	37.2	29.14	0.73	10.63
16*	10/9/2014	15:05	1.43	11	47.3	930.9	31.3	27.49	0.57	18.81
16	10/9/2014	15:57	1.61	13	45	932.4	32.3	27.49	0.57	18.81
16	10/9/2014	16:03	1.60	13	40.4	931.8	33.3	27.49	0.57	18.81
16*	10/9/2014	16:01	1.68	15	38.7	932	34.3	27.49	0.57	18.81
16	10/9/2014	16:16	1.61	17	37.1	932	35.3	27.49	0.57	18.81
16	10/9/2014	16:22	E60	15	37.3	932	36.3	27.49	0.57	18.81
9	10/23/2014	7:15	0.64	64	23.9	939	28.5	22.55	0.41	9.51
9	10/24/2014	7:03	0.68	54	26.3	939.5	28.5	22.55	0.41	9.51
9	10/25/2014	7:04	0.65	48	28	939	28.5	22.55	0.41	9.51
9*	10/26/2014	7:45	0.70	45	29.3	940	28.5	22.55	0.41	9.51
9	10/27/2014	7:52	0.59	45	30.4	940.4	28.5	22.55	0.41	9.51
9*	10/28/2014	8	0.68	45	30.4	940.4	28.5	22.55	0.41	9.51

10	10/29/2014	8:14	0.93	40	34	940.4	26.5	23.94	0.49	9.95
10	10/30/2014	8:31	0.93	28	40.7	940.3	26.5	23.94	0.49	9.95
10*	10/31/2014	8:29	1.07	22	42.9	940	26.5	23.94	0.49	9.95
10	11/1/2014	8:37	0.86	14	44.3	940	26.5	23.94	0.49	9.95
10	11/2/2014	8:46	0.86	25	41.5	940	26.5	23.94	0.49	9.95
6	11/3/2014	8:59	0.61	22	42.4	940.2	25.2	23.32	0.34	10.69
6	11/4/2014	9:07	0.69	29	39.3	940.5	25.2	23.32	0.34	10.69
6	11/5/2014	9:16	0.64	25	40.5	940.5	25.2	23.32	0.34	10.69
6	11/6/2014	9:31	0.58	23.2	37.3	940.6	25.2	23.32	0.34	10.69
6*	11/7/2014	9:36	0.51	33	36.7	940.6	25.2	23.32	0.34	10.69
11	11/8/2014	9:46	0.75	31	37.9	940.4	30.2	25.1	0.80	9.68
11	11/9/2014	9:53	0.98	26	41.6	940.4	30.2	25.1	0.80	9.68
11	11/10/2014	9:59	0.93	14	44.8	940.4	30.2	25.1	0.80	9.68
11*	11/11/2014	10:07	1.09	13	45.7	940.4	30.2	25.1	0.80	9.68
11	11/12/2014	10:13	0.86	11	47.3	940	30.2	25.1	0.80	9.68

Note: *Data not used in statistical analysis.

TABLE 3: CO EMISSIONS$_2$, AVERAGE ATMOSPHERIC PARAMETERS AND PHYSICO-CHEMICAL PARAMETERS OF THE SOIL IN PLOT 15.

Point	Date	Schedule	Emission (μmol CO2 ITT^2 s^1 -)	Umi, Air (%)	Ar	Air temp. Air (° C)	P Atm. (IiPa)	Humid. Soil (%)	Soil Temp. Soil (° C)	Thermal Condition (Wm^1 K')^1	C/N
4	11/10/2014	13:02	3.86	53.00		28	940.4	53.8	23.02	0.299	8.19
4	11/10/2014	13:27	2.54	56.00		28	940.4	53.8	23.02	0.299	8.19
4	11/10/2014	13:35	2.38	64.00		29.5	940.3	53.8	23.02	0.299	8.19
4	11/10/2014	13:42	2.30	54.00		30.1	940.3	53.8	23.02	0.299	8.19
4	11/10/2014	13:52	2.08	54.00		29.5	940.2	53.8	23.02	0.299	8.19
3	11/10/2014	14:08	1.59	47.00		30.5	940.3	47.1	23.43	0.289	9.66
3	11/10/2014	14:19	1.68	49.00		30.1	940.2	47.1	23.43	0.289	9.66
3	11/10/2014	14:34	1.56	56.00		29.5	940.3	47.1	23.43	0.289	9.66
3	11/10/2014	14:39	1.55	54.00		29.3	940.1	47.1	23.43	0.289	9.66
3	11/10/2014	14:48	1.56	66.00		28.6	939.9	47.1	23.43	0.289	9.66
2	11/10/2014	15:02	1.47	57.00		28.7	940	50.5	23.3	0.449	8.67
2	11/10/2014	15:08	1.03	62.00		28.7	939.7	50.5	23.3	0.449	8.67
2	11/10/2014	15:18	1.11	64.00		28.4	939.6	50.5	23.3	0.449	8.67
2	11/10/2014	15:25	0.92	66.00		28.1	939.6	50.5	23.3	0.449	8.67
2	11/10/2014	15:36	1.19	62.00		28.3	939.4	50.5	23.3	0.449	8.67
1	11/10/2014	15:53	1.32	61.00		28.7	939.5	48.8	22.95	0.32	8.42
1	10/11/1014	16:01	1.09	66.00		28.7	939.5	48.8	22.95	0.32	8.42
1	11/10/2014	16:07	1.25	65.00		28.2	939.3	48.8	22.95	0.32	8.42
1	11/10/2014	16:15	1.39	58.00		27.9	939.2	48.8	22.95	0.32	8.42
8	11/11/2014	14	0.85	47.00		27	940.2	43.8	21.42	0.48	11.39
8	11/11/2014	14:01	0.76	49.00		27	940.3	43.8	21.42	0.48	11.39
Point	Date	Schedule	Emission (μmol CO2 m^2 s^1 ·)	Umi, Air (%)	Ar	Air temp. Air (° C)	P Atm. (hPa)	Humid. Soil (%)	Soil Temp. Soil (° C)	Tèrni. cond. (Wm^1 K')^1	C/N

34

Point	Date	Schedule	Emission (µmol CO2 m^2 s^1 ·)	Umi, Air (%)	Air temp. Air (°C)	P Atm (hPa)	Humid. Soil (%)	Soil Temp. Soil (°C)	Cond. Tèrni (Wm^1 K')^1	C/N
8	11/11/2014	14:02	0.61	48.00	27.1	940.2	43.8	21.42	0.48	11.39
8	11/11/2014	7:12	0.89	48.00	27.2	940.3	43.8	21.42	0.48	11.39
7	2/3/2015	15	3.04	80.00	26	939.5	65.6	22.51	0.58	10.61
7	2/3/2015	15:15	2.92	78.00	26.2	939.5	65.6	22.51	0.58	10.61
7	2/3/2015	15:22	2.76	78.00	26.5	939.3	65.6	22.51	0.58	10.61
7	2/3/2015	15:03	1.97	77.00	26	939.2	65.6	22.51	0.58	10.61
6	2/3/2015	15:04	1.75	70.00	26	939.7	60.6	22.48	0.55	10.81
6	2/3/2015	16	2.57	72.00	25.8	939.6	60.6	22.48	0.55	10.81
6	2/3/2015	16:01	1.23	70.00	25.8	939.6	60.6	22.48	0.55	10.81
6	2/3/2015	16:19	3.35	68.00	26	939.4	60.6	22.48	0.55	10.81
6	2/3/2015	16:03	2.73	68.00	26	939.5	60.6	22.48	0.55	10.81
5	2/3/2015	16:45	2.07	65.00	25.5	940.2	53.8	22.85	0.72	10.46
5	2/3/2015	17	2.57	60.00	25.5	940.3	53.8	22.85	0.72	10.46
5	2/3/2015	17:15	2.86	60.00	25.3	939.5	53.8	22.85	0.72	10.46
5	2/3/2015	17:03	3.02	60.00	25.3	939.5	53.8	22.85	0.72	10.46
9	24/03/2015	14:37	1.59	81.00	25.2	945.4	57.2	21.17	0.66	8.77
9	24/03/2015	14:45	1.95	84.00	25.2	945.4	57.2	21.17	0.66	8.77
9	24/03/2015	14:52	1.99	81.00	25.4	945.2	57.2	21.17	0.66	8.77
9	24/03/2015	15:01	1.98	80.00	25.2	945.1	57.2	21.17	0.66	8.77
13	17/04/2015	10:23	1.64	86.00	23.2	945.9	57.2	21.42	0.48	9.66
14	17/04/2015	10:04	1.73	88.00	23.3	945.6	53.8	21.42	0.48	9.66
15	17/04/2015	10:55	2.59	89.00	23.5	945.3	63.9	21.42	0.48	9.66
16	17/04/2015	11:05	2.14	89.00	24.2	945.2	50.5	21.42	0.48	9.66
Point	Date	Schedule	Emission (µmol CO2 m^2 s^1 ·)	Umi, Air (%)	Air temp. Air (°C)	P Atm (hPa)	Humid. Soil (%)	Soil Temp. Soil (°C)	Cond. Tèrni (Wm^1 K')^1	C/N
10	17/04/2015	11:02	2.49	88.00	24.7	945.0	67.3	21.42	0.48	9.66
17	13/05/2015	9:05	2.10	89.00	19.4	948.6	39.9	18	0.48	9.66
14	13/05/2015	09:34	2.40	77	21	949.3	70.0	18	0.48	9.66
И	13/05/2015	10:05	2.03	88.00	19.6	948.8	39.9	18	0.48	9.66
12	13/05/2015	10:15	1.67	92.00	18.9	948.5	34.0	18	0.48	9.66

TABLE 4: DESCRIPTIVE STATISTICS OF THE PARAMETERS STUDIED IN THE PROJECT.

	Hoist	Emission (µmol CO2 m^2 s^1 ·)	Umi. Air (°C)	Air temp. Air (°C)	P Atm (IrPa)	Humid. Soil (%)	Soil Temp. Soil (°C)	Cond. Tèrni (Wnr^1 K')^1	C/N	Schedule
Media		E38	23.97	38.22	937.14	32.10	25.20	0.74	11,40	12.36
Max.		2.59	64.00	50.20	940.80	48.19	35.30	1.07	18,81	17.18
Min	15	0.51	11.00	23.70	930.90	21.65	18.31	0.34	8,57	7.25
DV		0.54	12.66	7.51	3.86	7.10	4.50	0.25	2,55	3.14
CV		39.13	52.82	19.65	0.41	22.12	17.86	33.78	22,37	25.40
Median		1.52	22.00	37.80	940.00	30.49	23.94	0.80	10,63	11.67
Media		1.92	67.84	26.36	941.58	53.34	22.12	0.48	9,64	14.17
Max.	23	3.86	92.00	30.50	949.30	70.00	23.43	0.72	11,39	17.50
Min		0.61	47.00	18.90	939.20	33.99	18.00	0.29	8,19	7.20

DV		0.73	13.39	2.70	3.04	7.77	1.44	0.13	1,01	2.31
CV		38.02	19.74	10.24	0.32	14.57	6.51	27.08	10,48	16.30
Median		1.95	66.00	26.20	940.20	53.85	22.51	0.48	9,66	14.80
Media		1.63	46.08	32.18	939.45	42.54	23.56	0.60	10,45	13.16
Max.		3.86	92.00	48.30	949.30	70.00	35.30	1.07	18,81	17.50
Min.	Total	0.51	11.00	18.90	931.50	21.65	18.00	0.29	8,19	7.20
CV		31.29	23.87	58.73	99.15	50.89	76.40	48.33	78,37	54.71
DV		0.70	25.47	7.97	4.03	13.07	3.50	0.24	2,10	3.00
Median		1.61	48.00	29.00	940.05	41.72	23.02	0.55	10,01	14.31

With regard to CO_2 fluxes from soil carbon emissions, it was observed that in plot 15 emissions were slightly lower than those observed in plot 23, respectively between 0.51 and 2.59 µmol CO_2 m^{-2} s^{-1}, with an average of 1,38 µmol CO_2 m^{-2} s^{-1}, a standard deviation of 0.54 and a median of 1.52 µmol CO_2 m^{-2} s^{-1}, against emissions ranging from 0.61 to 3.86 µmol $CO2$ m^{-2} s^{-1}, with an average of 1.92 µmol CO_2 m s^{-2-1}, a standard deviation of 0.73 and a median of 1.95 µmol CO_2 m s^{-2-1} (Table 3). Figure 18 shows these differences.

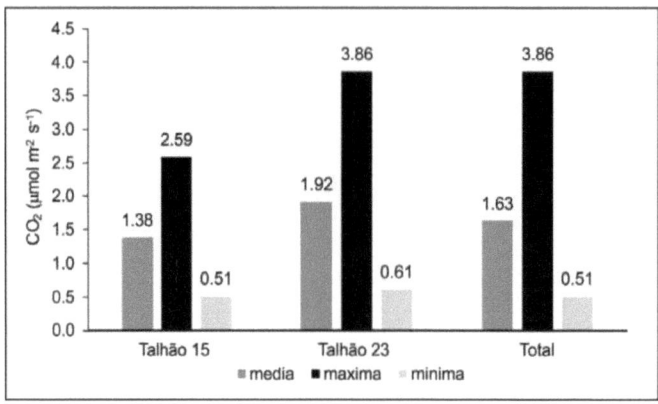

FIGURE 18: CO2 EMISSIONS.

Due to the data collection period and the shading provided by the trees, the temperature variations measured in plot 15 are greater than in plot 23, respectively between 23.7°C to 50.2°C (average 38.3°C) and 18.9°C to 30.5°C (average 26.4°C). The soil temperature shows similar behavior, with variations from 18.3°C to 35.3°C (average of 25.2°C) in plot 15, and between 18.0°C and 23.4°C (average of 22.1°C) (Table 3 and Figure 19).

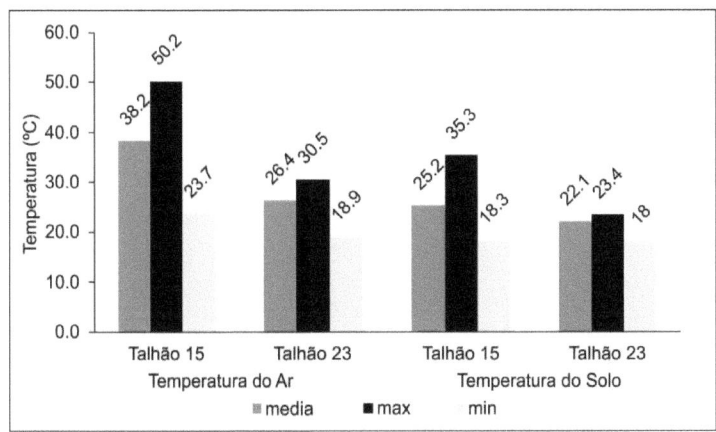

Figure 19: Soil and air temperature during the collection period.

Relative humidity in plot 15 ranged from 11.0% to 64.0% (average 23.9%), while in plot 23, due to the presence of vegetation, humidity ranged from 47.0% to 92.0% (average 67.8%) (Figure 20). Small variations in atmospheric pressure were observed, between 930.9hPa and 940.8hPa in plot 15, and between 939.2 hPa and 949.3 hPa in plot 23.

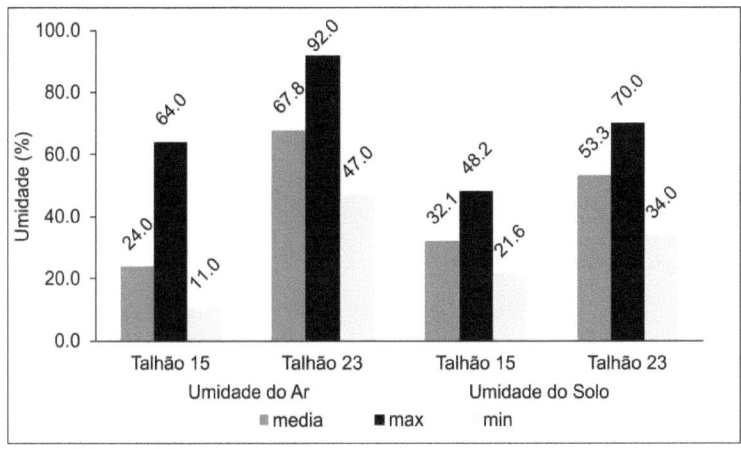

FIGURE 20: SOIL AND AIR HUMIDITY DURING THE COLLECTION PERIOD.

As a result of the presence of vegetation cover on the ground, the soil's physical parameters show some differences between the areas. Soil moisture in plot 15 ranged from 21.7% to a maximum of 48.2% (average 32.1%), while in plot 23 it varied from 34.0% to 70.0% (average 53.3%) (Figure 20). The thermal conductivity of plot 15 is higher than that of plot 23, ranging from 1.07 W m^{-1} K^{-1} to 0.34 W m K^{-1-1} (average of 0.74 W m^{-1} K^{-1}), and from 0.72 W m^{-1} K^{-1} to 0.29 W m K^{-1-1} (average of 0.48 W m^{-1} K^{-1}), respectively (Figure 21).

37

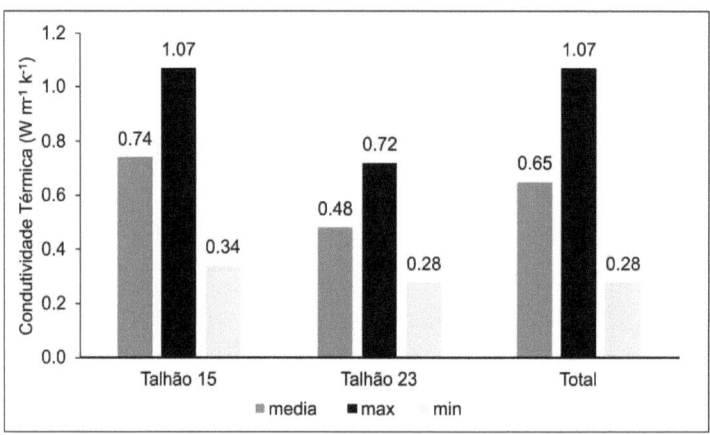

FIGURE 21: THERMAL CONDUCTIVITY DURING THE COLLECTION PERIOD.

The C/N ratio showed higher values in plot 15, indicating a lower presence of carbon in the soil, ranging from 18.8 to 8.6 (average 11.44) in plot 15, and from 11.4 to 8.2 (average 9.6) in plot 23 (Figure 22).

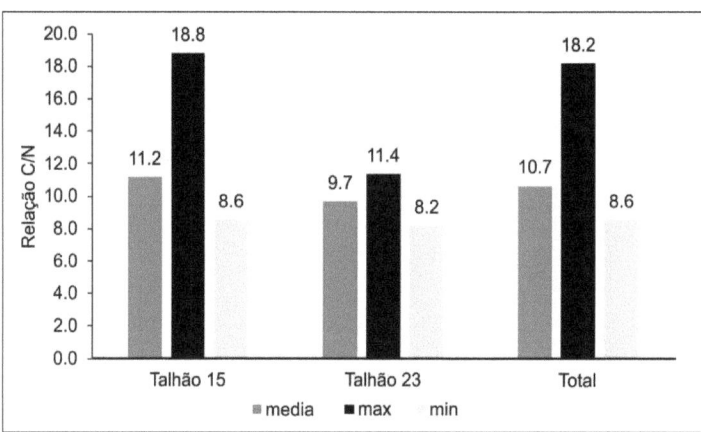

FIGURE 22: MEASURED C/N RATIOS.

5.3 Evaluation of Daily Emissions Fluctuations

Various studies on CO_2 emissions and soil respiration indicate that emissions fluctuate on a daily basis (LA SCALA et al. 2000; TEIXEIRA et al., 2011; EPRON, 2014; BICALHO et al., 2014). Field information was collected on different days and at different times of the day, and in order to understand these oscillations in the study area, the average daily values of the measured emissions were calculated, as well as the parameters soil moisture and air temperature, variables that are significantly correlated with soil respiration (DIAS, 2006; EPRON et al., 2006; OHASHI AND GYOKUSEN, 2007).

Date	CO emissions$_2$ (μmol CO_2 m^{-2} s)$^{-1}$		Soil moisture (%)		Air temperature (°C)		n
	Media	cv*(%)	Media	CV(%)	Media	CV(%)	
07/09/14	1,98	16,67	33,32	23,62	30,38	12,94	16
08/09/14	1,73	13,29	31,42	26,88	38,08	17,04	13
09/09/14	1,32	24,24	36,06	17,25	45,1	7,82	21
23/09/14	0,77	21,46	27,63	6,75	37,39	18,02	21
10/10/14	1,68	40,9	50,13	5,06	28,88	2,67	19
11/10/14	0,78	14,04	43,75	0	27,08	0,31	4
03/02/15	2,53	22,89	60,06	7,72	25,84	1,32	13
24/03/15	1,88	8,82	57,21	0	25,25	0,34	4
17/04/15	2,12	18,16	58,56	10,66	23,78	2,43	5
13/05/15	2,05	12,71	45,94	30,69	19,73	3,95	4

obs:* CV- Coefficient of Variation

The average daily CO_2 emissions from the soil ranged from 0.77 to 1.98 μmol CO_2 m^{-2} s^{-1} for plot 15 (reforestation in the growth stage) and from 0.78 to 2.53 μmol CO_2 m^{-2} s^{-1} for plot 23 (established reforestation), showing the higher emission rates for the area that had already been reforested, as previously mentioned.

the coefficient of variation values were between 8% and 40%, low values compared to those found by other authors (BiCALHo et al., 2014) in the state of sao Paulo, it should be taken into account when analyzing these values that in this project each point was measured more than once, and on some days few measurements were taken.

When comparing the average soil moisture values and the emission rates (Table 5), it can be seen that there is a correlation between the values, with an increase in moisture corresponding to an increase in emissions. However, this correlation is not statistically significant ($r=0.60$, $p<0.06$), mainly due to the small number of samples.

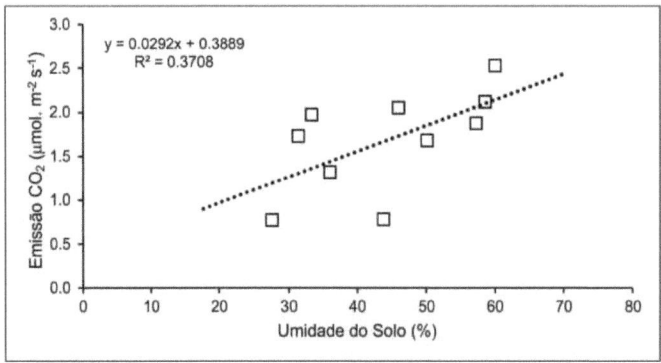

FIGURE 23: RELATIONSHIP BETWEEN DAILY AVERAGE CO2 AND SOIL MOISTURE.

When compared with the measured temperatures, there is a negative linear correlation between the average daily CO_2 emission and the average daily temperature (Figure 24), again not statistically significant ($r=-0.49$, $p<0.2$). Although not significant, the negative correlation can be explained by the measurement of higher emission rates (Figure 18) in the restored forest area (plot 23), where temperatures are lower and more homogeneous (Figure 19).

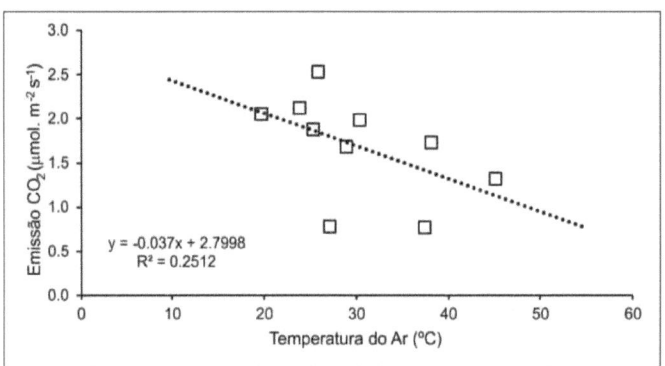

FIGURE 24: RELATIONSHIP BETWEEN DAY AVERAGES OF EMISSIONS AND AIR TEMPERATURE.

5.4 CO2 Emissions and Environmental Variables

In order to better study the relationships between the variables measured in the project, a correlation matrix was drawn up for the data collected in the newly reforested area so that the correlations between the independent variables could also be assessed. Table 6 shows the correlation matrix for the data collected in plot 15.

TABLE 6: CORRELATION MATRIX OF THE AREA PLANTED IN 2014. VALUES MARKED ASTERISCO $p<0.05$

I saw Issue	V2 Humid. Air	V3 Air Temp	V4 Pressure	V₅ U do soil	V6 T.Solo	V7Cond. Term.	V8 C/N	V9 Timetable

V1 saíd	1,00	-	-	-	-	-	-	-	-
V2	-0,11	1,00	-	-	-	-	-	-	-
V3	-0,21	-0,84*	1,00	-	-	-	-	-	-
V4	0,02	0,61*	-0,47*	1,00	-	-	-	-	-
V5	0,18	-0,09	-0,10	" -****z -0,46	1,00	-	-	-	-
V6	-0,02	-0,62*	0,75*	-0,52*	-0,24*	1,00	-	-	-
V7	0,56*	-0,16	-0,16	-0,20	0,57*	-0,26*	1,00	-	-
V8	0,28**	-0,37*	0,29*	-0,44*	-0,08	0,46*	-0,16	1,00	-
V9	0,33*	-0,74*	0,47*	-0,79*	0,31*	0,58*	0,26*	0,44*	1,00

Soil CO_2 emission rates showed a significant linear correlation with three of the variables studied (Table 6): thermal conductivity ($r=0.56$, $p<0.0001$), C/N ratio ($r=0.28$, $p<0.05$) and time of day ($r=0.33$, $p<0.05$).

Air temperature showed a non-significant negative correlation ($r=-0.21$, $p<0.1$) with emissions, as did the correlation between daily soil respiration and temperature, in plot 15 this is due to the fact that extreme hot temperatures occurred during the collection period, which ends up inhibiting bacterial activity (KANG et al., 2003).

Soil moisture showed a non-significant linear correlation ($r=0.18$, $p<0.2$) with respiration. The significant positive linear correlation ($r=0.57$, $p<0.0001$) between thermal conductivity and soil moisture, as well as the correlation between thermal conductivity and emissions may be indicating the effect of moisture on soil respiration.

Although the time of day does not directly influence emissions, it may be representing the influence of environmental variables that have a correlation, the variable shows a linear correlation.

with air temperature ($r=0.47$, $p<0.0001$), soil temperature ($r=0.58$, $p<0.0001$) and soil moisture ($r=0.31$, $p<0.01$).

We know that the site has just been reforested, so emissions have little influence from the respiration of plant roots. In this case, the amount of carbon in the soil can be a determining factor for the amount of CO_2 emitted, since the C/N ratio showed a positive linear correlation with emissions.

The same procedure was carried out for plot 23, an area reforested in 1918 for the same purpose (Table 7).

TABLE 7: CORRELATION MATRIX OF THE AREA PLANTED IN 1918 (PLOT 23). VALUES MARKED WITH ASTERISK $P<0.05$

V1 Issue	V2 Humid. Air	V3 Air Temp	V4 Pressure	V5 U do soil	V6T.Solo	V7Cond. Term.	V8 C/N	V9 Schedule

V1	1,00	-	-	-	-	-	-	-	-
V2	0,28*	1,00	-	-	-	-	-	-	-
V3	-0,24	-0,77*	1,00	-	-	-	-	-	-
V4	0,07	0,72*	-0,81*	1,00	-	-	-	-	-
V5	0,55*	0,34*	-0,05	-0,06	1,00	-	-	-	-
V6	-0,02	-0,60*	0,89*	-0,88*	0,14	1,00	-	-	-
V7	0,27	0,41*	-0,51*	0,18	0,38*	-0,26	1,00	-	-
V8	0,04	-0,03	-0,28	-0,12	0,11	-0,17	0,48*	1,00	-
V9	0,08	-0,36*	0,51*	-0,71*	0,23	0,68*	0,22	-0,01	1,00

Talhao 23 showed a significant linear correlation (Table 7) between CO_2 emissions and soil humidity ($r=0.55$, $p<0.0001$), and air humidity ($r=0.28$, $p<0.05$).

The correlation between soil moisture and CO_2 respiration has already been shown when analyzing daily moisture fluctuations. As moisture increases, there is an increase in the degradation activities of M.O. by microorganisms (KUTsCH et al., 2010).

Air humidity in plot 23 shows a significant correlation with soil humidity ($r=0.34$, $p<0.05$), indicating that higher air humidity is linked to rainfall events. As with plot 15, air temperature shows a negative linear relationship ($r=-0.24$, $p<0.11$) which is not significant with emissions.

When the correlation matrix is evaluated with all the data collected (Table 8) from the two areas together, it can be seen that CO_2 emissions show a significant positive correlation with the following parameters: air humidity ($r=0.40$, $p<0.0001$), atmospheric pressure ($r=0.25$, $p<0.05$), soil humidity ($r=0.55$, $p<0.0001$) and time of day ($r=0.33$, $p<0.01$) and significant negative correlation with air temperature ($r=-0.41$, $p<0.0001$).

TABLE 8: CORRELATION MATRIX OF TOTAL PROJECT DATA. VALUES MARKED ASTERISCO $p<0.05$

	V1 Issue	V2 Humid. Air	V3 Air Temp	V4 Pressure	V5 U of soil	V6 T.Solo	V7 Cond. Term.	V8 C/N	V9 Schedule
V1	1,00	-	-	-	-	-	-	-	-
V2	0,40*	1,00	-	-	-	-	-	-	-
V3	-0,41*	-0,89*	1,00	-	-	-	-	-	-
V4	0,25*	0,74*	-0,67*	1,00	-	-	-	-	-
V5	0,55*	0,74*	-0,63*	0,31*	1,00	-	-	-	-
V6	-0,17	-0,62*	0,79*	-0,66*	-0,38*	1,00	-	-	-
V7	0,09	-0,44*	0,28*	-0,36	-0,21*	0,05	1,00	-	-
V8	-0,01	-0,44*	0,40*	-0,46	-0,31*	0,48*	0,18	1,00	-
V9	0,33*	0,00	0,04*	-0,44	0,42*	0,35*	0,05	0,15	1,00

Comparing the data collected, it can be seen that the highest CO_2 emissions (Figure 18) were recorded in plot 15, as well as the highest soil and air humidity (Figure 20), C/N ratio (Figure 22) and atmospheric pressure (Table 4), while the highest soil and air temperature (Figure 19) and thermal conductivity (Figure 21) were recorded in plot 23.

This explains why, when we analyze the data together, we see a stronger correlation between air humidity and emissions than when we evaluate the individual areas. In restored forest areas, there are higher air humidities (Figure 20) and higher CO_2 emissions when compared to the newly reforested area, which is why the total data showed this correlation (Table 8), which was already evident in the area planted in 1918 (Table 7).

Soil temperature showed a significant negative correlation with soil respiration (Table 8), a trend already observed in the analysis of the average daily data (Figure 23), and in the individual analysis of each of the plots (Tables 6 and 7). This is due to the fact that the lowest air temperatures are found in forest areas, because of the microclimate created by the vegetation.

The relationship between soil moisture and respiration (Table 8) was similar to the value found for the area reforested in 1918 (Table 7), showing that moisture is an important controller of emissions in both newly reforested and restored areas, with the strongest correlation found in our set of variables.

5.5   Multiple linear regression

From the analysis of the correlation matrix with all the project data (Table 8), it can be seen that there are several variables correlated with CO_2 emissions, but none of them is capable of satisfactorily predicting the CO_2 emission rate from soil respiration. Multiple linear regression is the appropriate statistical tool for predicting a dependent variable when it correlates with several independent variables.

Due to the relationship between the number of independent variables and the number of samples, the development of a regression model for each of the areas separately would lead to problems of overfitting (HAIR Jr et al., 2009), and the development of a single model is recommended, since both areas are on the same type of soil and climatic regime.

In order to achieve the aim of the multiple linear regression, which is to estimate a general model for predicting CO_2 in reforested areas of the Atlantic Forest, it was necessary to standardize the number of samples for both plots. The procedure was the same as that used previously (Table 8), using the average of the values calculated at each point (Table 2) to eliminate the measurements with the greatest deviations from the average (Appendix 1).

To assess the ability of the selected independent variables to predict CO_2 emissions from the soil, a multiple linear regression equation was estimated using the *Stata: Data Analysis and Statistical Software* software using the data in Annex 1. Table 9 shows the results of the multiple regression.

TABLE 9: MULTIPLE LINEAR REGRESSION WITH ALL THE DATA COLLECTED.

Num. Obser.=98				SS	df	MS
F(8, 89)= 12,68			Regression	25,94612	8	3,24326526
Prob >F = 0	Root MSE = 0.50581		residual	22,76997	89	0,25584241
R² = 0.53	R² adjusted = 0.49		total	48,7161	97	0,5022278
Variable	Coef.	Standard Error	t	P>|t|	(Interv. Conf. 95%)	
Temp, air	-0,64313	0,0207775	-3,10	0,003	-0,1056	-0,0230282
Humid, Air	-0,01890	0,0077692	-2,43	0,017	-0,0343	0,0034645
Temp, soil	0,09957	0,0328228	3,03	0,003	-0,0343	0,1647882
Umid, Solo	0,03462	0,008346	4,15	0,000	0,0180	0,512001
Pressure	0,11100	0,0264199	4,20	0,000	0,0585	1,634966
Cod. Term,	0,89699	0,2648216	3,39	0,001	0,3708	1,423190
C/N	0,05652	0,0293649	1,92	0,057	-0,0018	0,1148726
Schedule	0,03800	0,0281285	1,35	0,18	0,0179	0,0938250
cons	-105,1550	24,930580	-4,22	0,000	-154,6910	-55,618220

Analysis of the results indicates that it is possible to reject the hypothesis of no regression, i.e. the model is significant at a significance level of 0.05, since the F value (12.68) is greater than the critical value (Fs = 2.126) and the p-value = 0.0000 < 0.05, it can be concluded that at least one of the explanatory variables is related to Co2 emissions.

The model's ratio value is satisfactory ($R^2 = 0.53$), and represents the proportion of the variation in co2 emissions that is explained by the set of explanatory variables selected, as can be seen in figure 25, which shows the measured values *versus* the values calculated by the multiple linear regression. It can be seen that the calculated values (series 2) are better adjusted to the observed values (series 1) in plot 15 (recently reforested - points 1 to 49) than in plot 23 (reforested in 1918 - points 50 to 98).

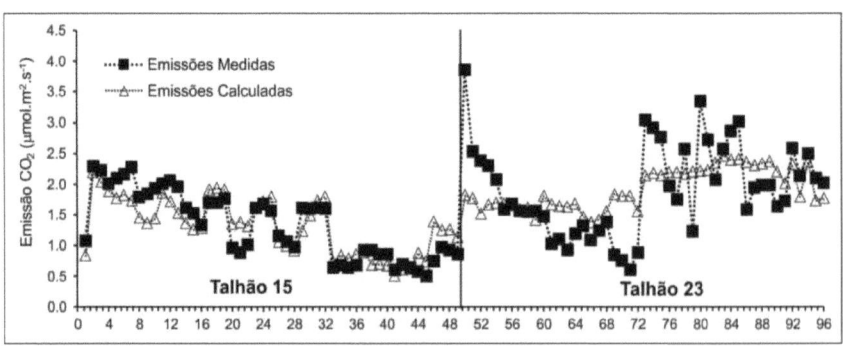

FIGURE 25: COMPARISON GRAPH OF MEASURED VALUES (IN BLUE) VS. CALCULATED VALUES (IN ORANGE).

Analysis of the residuals (Figure 26) shows that they do not have a constant variation, close to zero, increasing as a function of emissions, i.e. they show a tendency to move away, indicating the existence of heteroscedasticity, which is the violation of the statistical assumption that the variances of the error terms are equal (HAiR Jr. Et al., 2009).

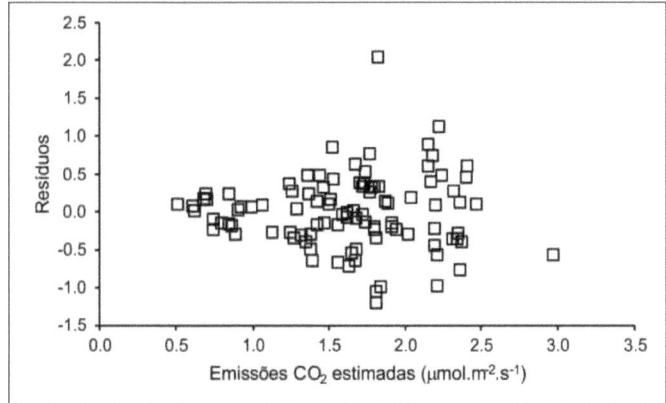

FIGURE 26: ADJUSTED VALUES *VERSUS* RESIDUALS. The DISTRIBUTION OF THE RESIDUALS SHOWS AN INCREASE IN DISPERSION AS EMISSIONS INCREASE, INDICATING HETEROSCEDASTICITY.

The existence of discrepant observations (*outliers*) in the recorded data meant that the model based on multiple linear regression presented two violations of statistical assumptions, which would not allow it to be validated.

To correct this problem, the Hubber-White standard error method (GREENE, 2008) was used, using *Stata* software, the results of which are shown in Table 10, indicating that heteroscedasticity was reduced, while the independent variable "C/N ratio" became significant at 5%, while the variable "air humidity" was significant at only 10%, and time remained insignificant.

TABLE 10: REGRESSION USING THE HUBBER-WHITE ROBUST ERROR METHOD (GREENDE, 2008).

Num. Obser.=98						
F(8, 89)= 25,69						
Prob>F= 0	Root MSE = 0.50581					
R^2 = 0.53						
Variable	Coef.	Standard Error	t	P>|t|	(Interv. conf. 95%)	
Air temp.	-0,64313	0,021435	-3,00	0,003	-0,10690	-0,021720
Humid. Air	-0,01890	0,009915	-1,91	0,060	-0,03860	0,000799
Soil	0,09957	0,024360	4,09	0,000	-0,05117	0,147972
Umi. Solo	0,03462	0,009724	3,56	0,000	0,01529	0,539382
Pressure	0,11100	0,021372	4,25	0,000	0,06853	0,153466
Cod. Ter.	0,89699	0,280551	3,20	0,002	0,33955	1,454443

45

C/N	0,05652	0,025240	2,24	0,028	0,00637	0,106676
Schedule	0,03800	0,024742	1,54	0,128	0,01116	0,087126
cons	-105,1550	19,83159	-5,30	0,000	-144,560	-65,74980

This method, therefore, did not adequately correct the problems observed in the initial regression, making it necessary to develop a third model, the robust regression (GREENE, 2008). In this type of regression, *outliers* are not included in the analysis, making it possible to solve the two problems encountered, heteroscedasticity and the existence of discrepant observations (Figure 26). Table 11 shows the results of this regression, indicating that all the variables are significantly important (p-value < 0.05).

TABLE 11. ROBUST REGRESSION RESULTS FOR THE TWO AREAS.

Num. Obser.=98						
F(8, 89)= 15,39						
Prob >F = 0						
Variable	Coef.	Standard Error	t	P>|t|	(95% confidence interval)	
Air temp.	-0,54158	0,018513	-2,93	0,004	-0,09094	-0,01737
Humid. Air	-0,01384	0,006923	-2,00	0,049	-0,02759	0,000081
T. soil	0,078809	0,029246	2,69	0,008	-0,02070	0,136919
Umi. Solo	0,024148	0,007436	3,25	0,002	0,009372	0,038924
Pressure	0,117122	0,023540	4,98	0,000	0,070348	0,163897
C. Term.	1,093309	0,235959	4,63	0,000	0,624464	1,562155
C/N	0,081705	0,026164	3,12	0,002	0,029717	0,133694
Timetable	0,060539	0,025063	2,42	0,018	0,010739	0,110338
cons	-111,238	22,21343	-5,01	0,000	-155,375	-67,10

It can be seen that this regression was able to reduce heteroscedasticity (Figure 27), reducing the distribution of residues for the highest emissions.

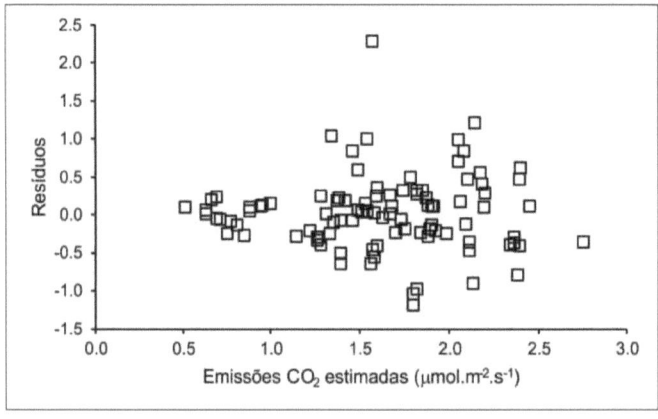

Figure 28 shows that for any of the regressions carried out, the models generated most accurately reproduce the emission values measured in field 15, while in field 23, which shows the greatest variability in the values measured in the field, none of the models are able to reproduce the extreme emissions (highest and lowest).

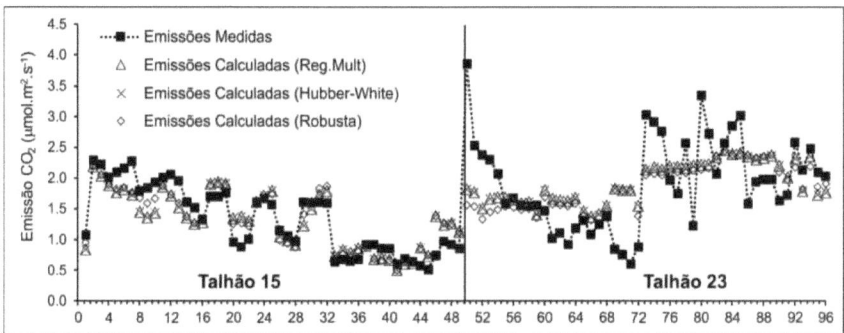

FIGURE 28: COMPARISON GRAPH OF OBSERVED VALUES X REGRESSIONS (OBSERVED - MEASURED values, CALCULATED - MULTIPLE LINEAR REGRESSION, HUBBER-WHITE - HUBBER-WHITE REGRESSION AND ROBUST - ROBUST REGRESSION).

This difference between the ability to predict emissions can be seen in figures 29 to 31, which show the coefficients of relationship between the measured and calculated values, while for all the measurements made the value of $R^2 = 0.51$, for plot 15 the linear relationship is $R^2 = 0.82$, and for plot 23 $R^2 = 0.19$.

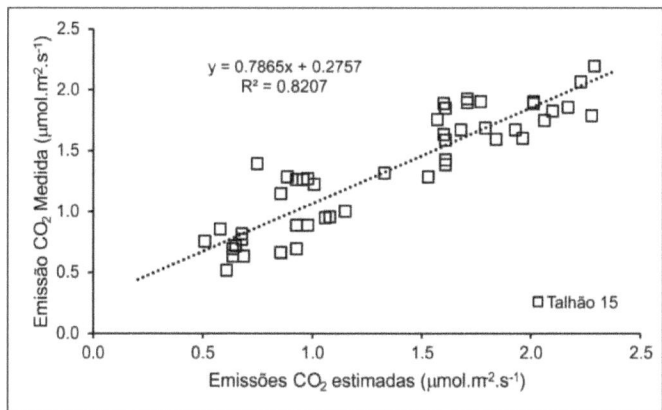

FIGURE 29: RELATIONSHIP BETWEEN OBSERVED EMISSIONS AND THOSE FORECAST BY THE ROBUSAR LYNEAR REGIME FOR PLOT 15.

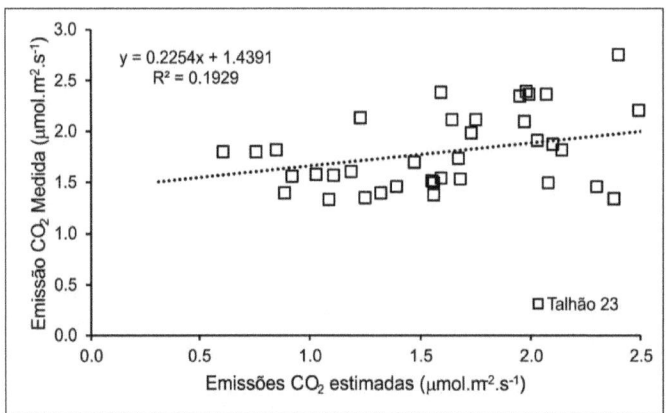

FIGURE 30: RELATIONSHIP BETWEEN OBSERVED EMISSIONS AND THOSE PREDICTED BY ROBUST LINEAR REGRESSION FOR BUTCHER BLOCK 23.

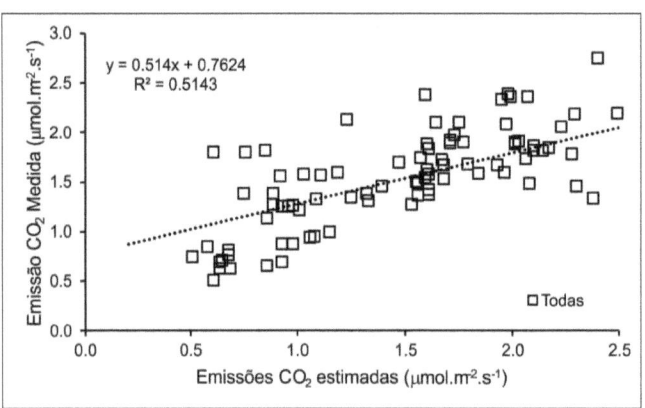

FIGURE 31: RELATIONSHIP BETWEEN EMISSIONS OBSERVED AND FORECASTED BY THE ROBUsTA LYNEAR REGIME FOR ALL MEASUREMENTS.

The independent variables selected are mainly climatic, such as temperature and humidity (Tables 2 and 3). There are indications of some determining factors for the lower respiration values in plot 15, but these were not evaluated in this project: the amount of shading of the soil by grass straw, as this can affect soil respiration rates, the amount of roots, related to the absence of tree individuals with an established root system, wind, direct solar radiation and the physical structure of the soil (KUTsH et al., 2010).

6. Final considerations and conclusions

The soil respiration values recorded during the execution of this project ranged from 0.51 μmol CO_2 m s^{-2-1} to 3.86 μmol CO_2 m s^{-2-1} (average of 1.63 μmol CO_2 m s^{-2-1}) (Figure 18), showing values similar to those obtained in experiments conducted in the interior of São Paulo on sugar cane crops (PANOSSSO et al, 2009; BRITO et al., 2010; BICALHO et al., 2014) and lower than those recorded in forest areas in the Amazon (NUNES, 2003; SOTTA et al., 2004; CHAMBERS et al., 2004; TRUMBORE et al., 2006; DIAS, 2006).

Within plot 15, reforested in 2014 (Figure 7), an average emission of 1.38 μmol CO_2 m s^{-2-1} was recorded. This plot was cultivated with eucalyptus from the beginning of the 20th century until 2003, when it was abandoned for 10 years and basically occupied by grass species, It was reforested this year and presents conditions closer to those of an area cultivated with sugar cane than those of a forest, due to the recently desiccated grass straw on the ground and the history of machinery traffic during the harvesting of the various eucalyptus cycles.

Plot 23, reforested in 1918 (Figure 8), had an average emission of 1.92 μmol CO_2 m s^{-2-1} (Figure 18), and is in an advanced state of regeneration, with a dense understorey, established trees in the forest canopy and re-established ecological functions.

The average soil respiration in plot 23 (1.92 μmol CO_2 m s^{-2-1}) was 31.25% higher than the average recorded in plot 15 (1.38 μmol CO_2 m s^{-2-1}). This difference is similar to the value attributed by some authors to "autotrophic" respiration, values between 40-70% (HANSON et al., 2000; BOND-LAMBERTY et al., 2004; SUBKE et al., 2006). For Davidson et al. (2002), given the same soil and climatic regime, the differences in emissions should be attributed to vegetation. However, considering the associated uncertainties, as well as the number of unmeasured variables, it is impossible to clearly distinguish their contribution.

CO_2 emissions showed a significant negative correlation with air temperature, whereas a positive correlation would be expected (RAICH & SCHLESINGER, 1992). This may be associated with the fact that at high temperatures microbial activity is reduced (KANG et al., 2003), and that during the collection period no low temperatures were recorded, but on the other hand high air temperatures were observed, extremes of up to 50°C, especially in plot 15 (Figure 7).

Soil temperature showed no significant correlation with emissions, as observed by other authors in the state of Sao Paulo (BICALHO et al., 2014). This can be attributed to the low temperature variations during the collection period, where plot 15 (Figure has the protection of straw and grass species, while plot 23 (Figure 8) has the protection of the forest canopy, which keeps the soil

49

temperature stable in both forest formations.

Soil moisture showed a significant correlation with soil respiration, a correlation observed by other authors, such as Dias (2006) and Shi et al. (2014), who justified the fact that microbial activity is regulated by moisture, due to the chemical reactions of decomposition of M.O. (KANG et al., 2003).

Thermal conductivity showed a significant positive correlation with CO_2 emission, especially in plot 15. This correlation between the thermal properties of the soil and respiration has already been shown in specific studies (NKONGOLO et al., 2010).

Air humidity showed a significant positive correlation with soil respiration, especially in plot 23, which has milder temperatures compared to plot 15. This correlation was not expected, since in temperate climates, a negative correlation between the variables has been observed (BILANDZIJA et al., 2014).

The time of collection showed a significant correlation with the soil respiration variable, especially in plot 15, and proved to be one of the independent variables to be used in predicting CO_2 emissions, with an important characteristic being the ease with which information can be collected. According to Singh and Gupta (1978), daily CO_2 oscillations can be explained by temperature fluctuations, which vary according to the time of day. On the other hand, we can see that the time of day correlates with several of the variables measured in the project.

As is well known, measuring CO_2 emissions from the soil depends on expensive equipment. The equipment developed at the Physics Department of Unesp in Rio Claro (MORENO, 2012) proved to be a viable alternative, at a lower cost, obtaining similar soil respiration values to other projects carried out in the state of Sao Paulo (PANOSSO et al., 2009; BRITO et al., 2010; BICALHO et al., 2014), and significant correlations with environmental variables suggested by specialized literature (LLOYD AND TAYLOR, 1994; DAVIDSON et al., 1998; EPRON et al., 2006; OHASHI AND GYOKUSEN, 2007; NKONGOLO et al., 2010; ALLAIRE et al., 2012), thus proving its effectiveness.

An analysis of the correlations between the independent variables and CO_2 emissions shows that none of them is capable of satisfactorily predicting soil respiration. The premise that soil respiration can be represented by linear relationships is not supported in the literature, even with the inclusion of various parameters (REICHSTEIN et al., 2002; 2005; DAVIDSON et al., 2006). This is possibly due to the number of factors that influence emissions and the difficulty of predicting extreme data.

However, the use of statistical methods, such as robust multiple linear regression, proved to be efficient in predicting emissions from newly forested areas, which can be explained by the existence of multicollinearity (tables 6 and 7). For example, soil thermal conductivity is a function of humidity, so in this case it is a co-variable. The existence of discrepant observations (*outliers*) in the recorded data was observed, presenting two violations of statistical assumptions, which would not allow it to be validated.

In order to correct the equation, Castellano et al. (2017) produced multiple linear regression models for the areas planted in 1918 and 2014, with a smaller number of random variables, considering only air temperature and humidity, atmospheric pressure, C/N ratio and soil moisture. The multiple correlation considering air temperature showed better results than the one considering soil temperature with one of the variables.

ACKNOWLEDGMENTS

The authors would like to thank CAPES for granting the first author a master's scholarship, FAPESP for project 00241-5/2012, the Physics Department of UNESP/Rio Claro for logistical support and the researchers who collaborated on the project: André Moraes Dejuste, Flavio Henrique Rodrigues, Leandro Xavier, Amauri Antônio Mengârio, Sâmia Maria Tauk Tornisielo.

7. BIBLIOGRAPHICAL REFERENCES

AYRES, et al. *BIOSTAT 5.0*. Belém: MCT - CNPq, 2007.

ALEXANDER, M. *Introducción a laMicrobiologia delSuelo*. Mexico: AGT Editor, 1980, 491 p.

ALLAIRE, S. E. et al. Multiscale spatial variability of CO2 emissions and correlations with physico-chemical soil properties. *Geoderma*, Amisterdam, n. 170, p. 251-260, 2012.

BAYER, C. *Soil organic matter dynamics in soil management systems*. 1996. 240 f. Thesis (Doctorate in Agronomy) - Federal University of Rio Grande do Sul, 1996.

BAYER, C. et al. Tillage and cropping system effects on soil humic acid characteristics as determined by electron spin resonance and fluorescence spectroscopies. *Geoderma*, Amsterdam, n. 105, p. 81-92, 2002.

BAYER, C. et al. Carbon sequestration in two Brazilian Cerrado soils under no-till. *Soil Tillage Research*, Amsterdam, n. 86, p. 237-245, 2006.

BAYER, C. et al. Soil carbon stabilization and mitigation of greenhouse gas emissions in conservation agriculture. In: KLAUBERG FILHO O.; MAFRA, A.L.; GATIBONI L.C. (ed.). *Tópicos em ciência do solo*. Viçosa: SBCS, 2011. p. 55-118.

BILANDZIJA, D.; ZGORELEC, Z.; KISIE, I. The Influence of Agroclimatic Factors on Soil CO2Emissions. *Collegium Antropologicum,* n. 38, p. 77-83, 2014.

BISCALHO, E.S. et al. Spatial variability structure of soil CO2 emission and soil attributes in a sugarcane area. *Agriculture Ecosystems & Environment*, Amsterdam, n. 189, p. 206-215, 2014.

BOLINDER, M. A.; ANGERS, D.A.; GIROUX, M. & LAVERDIERE, M.R. Estimating C inputs retained as soil organic matter from corn (Zea mays L.). *Plant Soil*, n. 215, p. 85-91, 1999.

BOND-LAMBERTY, B.; WANG, C. K.; GOWER, S. T. A global relationship between the heterotrophic and autotrophic components of soil respiration? *Global Change Biology,* n. 10, p.1756-66, 2004.

BRITO, L.F. et al. Soil CO2 emission of sugarcane field as affected by topography. *Scientia Agricola*, Piracicaba, n. 66, p. 77-83, .2009.

BRITO, L.F. et al. Spatial variability of soil CO2 emission of sugarcane field in different topography positions. *Bragantia,* Campinas, n. 69, p. 10-27, 2010.

CALIJURI, C. C.; CUNHA, D.G.F.; MOCCELIN, J. Ecological Fundamentals and Natural Cycles. In: CALIJURI, C.C.; CUNHA, D.G.F. *EngenhariaAmbiental Conceitos, Tecnologia e Gestão*. Rio de Janeiro: Elsiever, 2013. p. 131-154.

52

CAMPANILI, M; SCHAFFER, W. B. (Org.). *Mata Atlàntica: patrimònio nacional dos brasileiros (Biodiversity 34)*. Brasilia: Ministry of the Environment, 2010. p. 1-408.

CARDOSO, E.L. et al. Carbon and nitrogen stocks in soil under native forests and pastures in the pantanal biome. *Pesquisa agropecuària brasileira,* Brasilia, v. 45, n. 9, p. 1028-1035, 2010.

CASTELLANO, G. R. *Soil CO2 Emission in Restoration Areas in the Atlantic Forest*. 2015. 88 f. Dissertation (Master's Degree in Geosciences and Environment). Institute of Geosciences and Exact SciencesZUniversidade Estadual Paulista, Rio Claro, 2015.

CASTELLANO, G. R. et al. Quantification of soil CO_2 emission in two forested areas under different regeneration stages in Atlantic Forest. *Quimica Nova*, Sao Paulo, v.40, n.4, 2017

CHAMBERS, J. Q. et al. Respiration from a tropical forest ecosystem: portioning of sourcers and low carbon use efficiency. *Ecological Applications*, Washington, v.14, p. 72-88, 2004.

CHICOTA, R. *Field evaluation of a segmented TDR for soil moisture determination*. 2003. 100 f. Dissertation (Master's Degree in Agronomy) - University of São Paulo Luiz de Queiroz School of Agriculture, Piracicaba, 2003.

CHUNG, H.; GROVE, J.H.; SIX, J. Indications for soil carbon saturation in a temperate agroecosystem. *Soil Science Society American Journal, Madison*, v. 72, p.1132-1139, 2008.

DAVIDSON, E.A.; JANSSENS, I.A.; LUO, Y.Q. On the variability of respiration in terrestrial ecosystems: moving beyond Q10. *Global Change Biology*, v. 12, 154-164, 2006.

DAVIDSON, E. A. et al. Belowground carbon allocation in forest estimated from literfall and IRGA-based soil respiration measurements. *Agricultural and Forest Meteorology*, San Andreans, v. 113, p. 39-41, 2002.

DAVIDSON, E. A; BELK, E.; BOONE, R. D. Soil water content and temperature as independent or confounded factors controlling soil respiration in a temperature mixed hardwood forest. *Global Change Biology*, n. 4, p. 217-227. 1998.

DENEF, K.; SIX, J. Contributions of incorporated residue and living roots to aggregate- associated and microbial carbon in two soils with different clay mineralogy. *European Journal of Soil Science*, n. 57, p. 774-786, 2006.

DENMANM, K.L. et al. Couplings Between Changes in the Climate System and Biogeochemistry. In: SOLOMON, S. et al. (eds) *Climate Change 2007*: The Physical Science Basis. Contribution of Working Group I to the Fourth Assessment Report of the Intergovernmental Panel on Climate Change. UK and USA: Cambridge University, 2007

DIAS, J. D. *co2 flux from soil respiration in areas of native forest in Amazonia.* 2006. 87 f.

Dissertation (Master's Degree - Ecology of Agroecosystems) - University of São Paulo Luiz de Queiroz School of Agriculture. Piracicaba, 2006.

DIXON, R.K. et al. Carbon pools and flux of global forest ecosystems. *Science,* New York, v. 263, p. 185-190, 1994.

DUAH-YENTUMI, S.; RONN, R., CHRISTENSES, S. Nutrients limiting microbial growth in a tropical forest soil of Ghana under different management. *AppliedSoilEcology,* Amsterdam, v. 8, p. 19-24. 1998.

SÂO PAULO (state). Department of the Environment. *Edmundo Navarro de Andrade State Forest Management Plan.* CD ROOM: Forestry Institute, 2005.

INTERGOVERNMENTAL PANEL ON CLIMATE CHANGE. *The scientific bases - 2001.* Available at http://www.ipcc.ch/ipccreports/tar/wg1/. Accessed on August 3, 2014.

INTERGOVERNMENTAL PANEL CLIMATE CHANGE. *Climate Change 2001: Impacts, Adaptation and Vulnerability. Contribution of Working Group II to the Third Assessment Report of the Intergovernmental Panel on Climate Change.* UK and USA: Cambridge University Press, 2001.

EMBRAPA. National Soil Research Center. *Brazilian Soil Classification System.* 2 ed. Rio de Janeiro: Embrapa SPI, 2006. p. 306.

EPRON, D. et al. Soil CO_2 efflux in a beech forest: dependence on soil temperature and soil water content. *Annals of Forest Science,* Paris, v. 56, p. 221-6, 1999.

EPRON, D. et al. Spatial variation of soil respiration across a topographic gradient in a tropical rain forest in French Guiana. *Journal of Tropical Ecology,* Aberdeen, v. 22, p. 565-474, 2006.

FANG, C. et al. Soil CO_2efflux and its special variation in a Florida slash pine plantation. *Plant Soil,* v. 205, p. 135-146, 1998.

FAO - FOOD AND AGRICULTURE ORGANIZATION OF THE UNITED NATIONS. *State of the World's Forests 2001.* Rome: Food and Agriculture Organization. 2001. p. 181.

FERNANDES, T. J. G. *Contribution of reduced emission certificates (cers) to the economic viability of heveiculture.* 2003. 82 f. Thesis (Doctorate in Forestry Sciences) Universidade Federal de Viçosa, Viçosa. 2003.

FORSTER, H.W.; MELLO, A. C. G. Aerial root biomass in heterogeneous reforestation trees in the Paranapanema valley, SP. *Instituto Florestal - Série Registro,* Sao Paulo, n.31, p. 153-157, 2007.

FUENTES, J. P. et al. Microbial activity affected by lime in a long-term no-till soil. *Tillage Research,* Amsterdam, n. 88, p. 123- 131, 2006.

GALE, W.J.; CAMBARDELLA, C.A.; BAILEY, T.B. Surface residue and root-derived carbon in stable and unstable aggregates. *Soil Science Society of American Journal*, n. 64, p. 196-201, 2000.

GARDNER, W.H. Water content. In: KLUTE, A. (Ed.) *Methods of soil analysis I*: Physical and mineralogical methods. Madison: Soil Science Society of America, 1986. p. 493-544.

GARZELLA T. P. *Speedy reader reading automation and use in an irrigation management program.* 2011. 99 f Thesis (Doctorate) - University of Sao Paulo / Luiz de Queiroz College of Agriculture. 2011.

GRACE, J. Carbon Cycle. In: Simon Levin (Ed). *Encyclopedia of Biodiversity*, New York: Academic Press, 2001. p 69-629. v 1.

GREENE, W. H., *Econometric Analysis*. 6. ed. New Jersey: Prentice Hall, 2008. 1178 p.

GREGORICH, E.G.; ELLERT, B.H.; MONREAL, C.M. Turnover of soil organic matter and storage of corn residue carbon estimated from natural[13] C abundance. *Canadian Journal of Soil Science*, n. 75, p. 161-167, 1995.

GOLCHIN, A. et al. Soil structure and carbon cycling. *Australian Journal of Soil Research*, Victoria, n. 32, p. 1043-1068, 1994.

HAIR JR., J. F.; ANDERSON, R. E.; TATHAM, R. L.; BLACK, W. C. *Multivariate Data Analysis*. 5. ed. Porto Alegre: Bookman, 2005. 688 p.

HANSON, P. J. et al. Separating root and soil microbial contributions to soil respiration: a review of methods and observations. *Biogeochemistry*, Oregon, n. 48, p. 115-46, 2000.

HASSINK, J. The capacity of soils to preserve organic C and N by their association with clay and silt particles. *Plant Soil*, n. 191, p. 77-87, 1997.

HOGBERG, P.; NORDGREN, A.; BUCHMANN, N. Large-scale forest girdling shows that current photosynthesis drives soil respiration. *Nature*, n. 411, p. 789-92, 2001.

HORA R, C.; PRIMAVESI. O.; SOARES J.J. Contribution of liana leaves to litter production in a semideciduous seasonal forest fragment in Sao Carlos, SP. *Revista Brasileira de Botânica*, v.31, n.2, p.277-285, 2008.

JENKINSON, D.S. Soil organic matter: evolution. In: TERRON, P.U.; ROJO, C. (Ed) *Soil conditions and plant development according to Russell*. Madrid: Mundi Prensa, 1992. 500 p.

KANG, S. Y. et al. Topographic and climatic controls on soil respiration in six temperate mixed-hardwood forest slopes. Korea. *Global change Biology*, Oxon, v.9, n. 10, p. 1427-1437, 2003.

KELLER, M.; KAPLAN, W. A.; WOFSY, S. C. Emission of N_2O, CH_4 and CO_2 from tropical forest

55

soils. *Journal of Geophysical Research Atmospheres*, Washington, v.91, n.11, p.17911802, 1986.

KHOMIK, M.; ARAIN, M.A.; McCAUGHEY, J. H.; temporal and special variability of soil respiration in a boreal mixedwood forest. *Agricultural and Forest Meteorology*, Amsterdam, n.44, p. 244-256, 2006.

KJELDAHL, J. *Neue Methode zur Bestimmung des Stickstoffs in organischen Korpern*, Z. Anal. *Chem.*, v. 22, p. 366-382, 1883.

KLUTHCOUSKI, J.; AIDAR, H. Implementation, conduct and results obtained with the santa fé system. In: KLUTHCOUSKI, J.; STONE, L.F.; AIDAR, H. (Org.) *Crop-livestock integration.* Santo Antônio de Goiàs: Embrapa Arroz e Feijao, 2003. p.407-459.

KLUTHCOUSKI, J.; STONE, L.F. Performance of annual crops on Brachiaria straw. In: KLUTHCOUSKI, J.; STONE, L.F.; AIDAR, H. (eds). *Crop-livestock integration.* Santo Antônio de Goiàs: Embrapa Arroz e Feijao, 2003. p.500-522.

KOGEL-KNABNER, I. Analytical approaches for characterizing soil organic matter. *Geochem. Org.*, n. 31, p. 609-625, 2000.

KUNTORO, A.; WAHYU, A. The Effect of Deforestation on Regional Terrestrial Carbon Balance: A Case Study of Borneo Island. *Journal of International Development and Cooperation,* Japan, v. 15, p. 141-165, 2009.

KUTSCH. W. L.; BANH, M.; HEINEMEYER, A. *Soil Carbon Dynamic: an integrated methodology.* United Kingdom: Cambridge University Press, 2010, 298 p.

LA SCALA, Jr. N; PANOSSO A.R; PEREIRA G.T. Modelling short-term temporal changes of bare soil CO_2 emissions in a tropical agrosystem by using meteorological data. *Applied Soil Ecology*, v. 24, Amsterdam, p. 113-116, 2003.

LA SCALA, Jr. N. et al. Short term temporal changes in the spatial variability model of CO emissions from a Brasilian bare soil. *Soil Biology & Biochemistry*, Oxford, v.32, n.10, p. 14591462, 2000.

LEÓDIDO L.M. *Development of methods and means for the dynamic calibration of greenhouse gas transducers.* 2006. 106 f. Dissertation (Master's Degree) - Faculty of Technology/University of Brasilia - DF, Brasilia. 2006.

LI, Y.; LINDDSTROM, M.J. Evaluating soil quality-soil redistribution relationship on terraces and sep hillslope. *Soil Science Amstendars Journal.* v. 65, p. 1500 - 1508, 2001.

LLOYD, J.; TAYLOR, A. On the temperature dependence of soil respiration functional. *Ecology*, Oxford, v.8, n.3 p. 315-323, 1994.

LOVATO, T. et al. Carbon and nitrogen addition and its relationship with soil stocks and corn yield in management systems. *Revista Brasileira de Ciências do Solo*, n. 28, p.175-187, 2004.

MACHADO, F.B.; NARDY, A.J.R.; OLIVEIRA, M.A.F. Geology and petrological aspects of the Mesozoic intrusive rocks of the eastern edge of the Paranà Basin in the state of Sao Paulo. *Revista Brasileira de Geociências*, n. 37, p.64-80, 2007.

MCDOWELL, N.G. et al. Estimating CO_2 flux from snow packs at three sites in the Rock Mountains. *Tree Physiology*, n. 20, p.745-753, 2000.

MONTEIRO, C.A.F. - *Climate Dynamics and Rainfall in the State of São Paulo (Geographical Study in Atlas Form)*. Institute of Geography, USP, 1973.

MOREIRA R. M.; SILVA A. U. Leaf litter production and reforested area. *Revista Arvore*, Viçosa, v.28, n.1, p.49-59, 2004.

MORENO, L.X. *Development of a soil CO2 flux analysis system using the infrared radiation adsorption method*. 2012. 82 f. Dissertation (Master's Degree) - Institute of Geosciences and Exact Sciences/ "Julio de Mesquita Filho" Paulista State University, Rio Claro. 2012.

NCONGOLO, V. K. et al. Greenhouse gas fluxes and soil thermal properties in a pasture in central Missouri. *Journal of Environmental Sciences*, v. 22(7), p. 1029-1039.

NICOLOSO, R.S. *Soil organic carbon stabilization mechanisms in temperate and sub-tropical agroecosystems*. 2009. 108 f. Thesis (Doctorate) - Federal University of Santa Maria, Santa Maria. 2009.

NUNES, P. C. *Influence of soil CO2 efflux on forage production in an extensive pasture and an agrosilvopastoral system*. 2003. 68 f. Dissertation (master's degree in Tropical Agriculture Sciences) - Faculty of Agronomy and Veterinary Medicine/ Federal University of Mato Grosso, Cuiabà. 2003.

OADES, J.M.; GILLMAN, G.P.; UEHARA, G. Interactions of soil organic matter and variablecharge clays. In: COLEMAN, D.C.; OADES, J.M. & UEHARA, G. (Org.) *Dynamics of soil organic matter in tropical ecosystems*. Honolulu: Hawaii Press, 1989. p.69-95.

OADES, J.M.; WATERS, A.G. Aggregate hierarchy in soils. *Australian Journal of Soil Research*, Collingwood, v. 29, p.815-828, 1991.

ODUM , E. P. The strategy of ecosystem development. *Science*, n. 164, 262-70. 1969.

OHASHI, M., GYOKUSEN, K. Temporal chance in spatial variability of soil respiration on a slope of Japanese cedar (*Cryptomeria japonica* D. Don) forest. *Soil Biology and Biochemistry*, Oxford, n. 39, p. 1130- 1138, 2007.

PANOSSO, A.R. et al. Spatial and temporal variability of soil CO_2 emission in a sugarcane area under green and slash-and-burn managements. *Soil Tillage Research,* Amsterdam, n. 105, p. 275-282, 2009.

PANOSSO, A. R. et al. Soil CO_2 emission and its relation to soil properties in sugar cane areas under Slash-and-burn and Green Harvest. *Soil Tillage Research,* Amsterdam, n. 111, p. 190196, 2011.

PEIXOTO, M.F.S. *Physical, chemical and biological attributes as indicators of soil quality,* 2008.

PENTEADO, M.M.A. Tectonic implications in the genesis of the cuestas of the Rio Claro Basin (SP). In:(Org.)*Noticia Geomorfológica.* Campinas, vol. 15, no. 8, p. 19-41, 1968.

PENTEADO, M.M.A. Geomorphological study of the urban site of Rio Claro. In:(Org.) *Noticia Geomorfológica,* Campinas, n.° 42, p. 23-56, 1981.

PRIWITZER, T.; CAPULIAK, J.; BOSELA, M.; SCHWARS, M. Preliminary results of soil respiration in beech, spruce and grassy stands. *Lesnicky casopis - Forestry Journal,* Bratislava, n.59 (3), p 189-196, 2013.

REICHSTEIN, et al. Ecosystem respiration in two Mediterranean evergreen Holm Oak forests: drought effects and decomposition dynamics. *Functional Ecology,* v.16, p. 27-39, 2006.

REICHSTEIN, et al. On the separation of net ecosystem exchange into assimilation and ecosystem respiration: review and improved algorithm. *Global Change Biology,* v.11, p. 14241439, 2005.

RAICH, J. W; SCHLESINGER, W. H. The global carbon dioxide flux in soil respiration relationship to vegetation and climate. *Tellus,* Copenhagen, n. 44, p. 81-99, 1992.

RODRIGUES R. R. The vegetation of Piracicaba and the surrounding municipalities. *Circular tècnica IPEF,* Piracicaba, n. 189, p. 1-17, 1999.

ROSS, S. *Soil processes a systematic approach.* New York: Routledge, 1989, 444 p.

SABINE, C.L. et al. The oceanic sink for anthropogenic CO2, *Science,* v. 305, p. 367-371, 2004.

SABINO, C. V.; LAGE, V. L.; ALMEIDA, K. C. B. Use of robust statistical methods in environmental analysis. *Eng Sanit Ambiental,* special issue, p. 87-94, 2014.

SÂO PAULO (State). State Secretariat for the Environment - Biota Project - Sao Paulo. *Probio,* 1998.

SCHLESINGER, W. H. *Biogeochemistry: analysis of global change.* 2. ed. Oxon: Academic Press, 1997. 234 p.

SCHINDLBACHER, A. et al. Winter Soil respiration from an Austrian mountain forest. *Agricultural And Forest Metereology,* Amsterdam, n. 146, p. 205-215, 2007.

SHI W. Y. et al. Soil CO_2 emissions from five different types of land use on the semiarid Loess Plateau of China, with emphasis on the contribution of winter soil respiration. *Atmospheric Environment*, n.88, p.74-82, 2014.

SINGH, J. S.; GUPTA, S. R. Plant decomposition and soil respiration in terrestrial ecosystems. *Botanical Review*, New York, v.43, n.4, p.499-528, 1977.

SIX, J. et al. Stabilization mechanisms of soil organic matter: Implications for C-saturation of soils. *Plant Soil*, n. 241, p. 155-176, 2002.

SOTTA, E. D. *Flux of CO_2 between soil and atmosphere in a tropical rainforest in Central Amazonia*. 1998. 150 f. Dissertation (Master's Degree in Forestry Sciences) - National Institute for Amazonian Research, Manaus. 1998.

SOTTA, E. F. et al. Soil CO_2 efflux in a tropical forest in the central Amazon. *Global Change Biology*, Oxford, v.10, n.5, p. 601-617, 2004.

SIQUEIRA, J.O.; FRANCO, A.A. *Biotecnologia do solo: fundamentos e perspectivas*. Brasilia: MEC/ABEAS; Lavras: ESAL/FAEPE, 1988. 236 p.

STEWART, C.E. et al. Soil C saturation: linking concept and measurable C pools. *Soil Science Society of American Journal*, n. 72, p. 379-392, 2008

STEWART, C.E. et al. Soil carbon saturation: Implications for measurable carbon pool dynamics in long-term incubations. *Soil Biology & Biochemistry,* Oxford, n.41, p. 357-366, 2009.

SUBKE, J. A.; INGLIMA, I.; COTRUFO, M. F. Trends and methodological impacts in soil CO2 efflux partitioning: a meta-analytical review. *Global Change Biology*, n.12, p. 921-43, 2006.

TEIXEIRA, D.D.B. et al. Spatial variability of soil CO_2 emission in a sugarcane area characterized by secondary information. *Scientia Agricola,* Piracicaba, n. 70, p. 195-203, 2013.

THORNTWAITE, C.W.; MATHER, J.R. *The water balance.* Centerton, N.J.: The Laboratory of Climatology, 1981, 104 p.

TISDALL, J.M.; OADES, J.M. Organic matter and water-stable aggregates in soils. *Journal of Soil Science*, n. 33, p. 141-163, 1982.

TRUMBORE, S.E. et al. Seasonal variation in the soil respiration rate in coniferous forest soils. *Soils Biology & Biochemistry*, Oxford, v. 34, n.9, p. 1375-1379, 2002.

URQUIAGA, S. et al. Variations in carbon stocks and greenhouse gas emissions in soils of the tropical and subtropical regions of Brazil: A critical analysis. *Informe Agronomico*, n. 130, p.12-21, 2010.

59

RAZAFIMBELO, T.M. et al. Aggregate associated-C and physical protection in a tropical clayey soil under Malagasy conventional and no-tillage systems. *Soil & Tillage Research*, n. 98, p. 140-149, 2008.

VAN BAVEL, C. H. M. A soil aeration theory based on diffusion, *Soil Science,* n.72, p. 3346, 1951

VAN BAVEL, C. H. M. Gaseous diffusion and porosity in porous media. *Soil Science*, n. 73, p. 91-104, 1952.

VELOSO, H. P.; RANGEL FILHO, A. L. R.; LIMA, J. C. A. *Classification of Brazilian vegetation adapted to a universal system.* Rio de Janeiro: IBGE (Department of Natural Resources and Environmental Studies), 1991. 124 p.

VESTERDAK, L. et al. Carbon and nitrogen in forest floor and mineral soil under six common European tree species. *Forest Ecology and Management*, n.255, p 78-83, 2008.

WATSON, T.R.; NOBLE, R.I.; BOLIN, B.; RAVINDRANATH, N.H.; VERARDO, J.D.; DOKEN, J.D. *Land Use, Land Use Change, and Forestry.* A special report. Intergovernmental Panel on Climate Change. Cambridge, United Kingdom, Cambridge University Press. 2000.

YEOMANS, J.C. & BREMNER, J.M. A rapid and precise method for routine determination of organic carbon in soil. *Comm. Soil Sci. Plant Anal.*, 19:1467-1476, 1988.

ZALAMENA, J. *Impact of land use on the chemical and physical attributes of soils on the edge of the plateau - RS.* 2008. 79p. Dissertation (Master's Degree in Soil Sciences). University of Santa Maria - RS, Santa Maria, 2008.

ANNEX 01 - Data used to prepare the
multiple linear regression

Measure	Schedule	Issue	Umi. Air (°C)	Air temp. Air (°C)	P Atm. (hPa)	Humid. Soil (%)	Soil Temp. Soil (°C)	Cond. Tèrni, (W*nΓ^{-1} K)$^{-1}$	C/N
1	8,41	1,08	54	23,7	940,2	26,0	18,31	0,63	10,14
2	9,55	2,29	41	24,8	940,8	40,9	19,76	1,06	10,63
3	9,65	2,23	41	27,3	940,8	40,9	19,76	1,06	10,63
4	9,75	2,01	29	33,3	940,8	40,9	19,76	1,06	10,63
5	10,3	2,10	33	30,2	940,8	37,7	19,41	0,97	10,27
6	10,41	2,17	35	28,5	940,5	37,7	19,41	0,97	10,27
7	10,22	2,28	35	29,7	940,5	37,7	19,41	0,97	10,27
8	10,98	1,79	24	35,8	940,1	22,8	21,82	1,07	11,94
9	11,04	1,84	22	37,2	939,7	22,8	21,82	1,07	11,94
10	11,23	1,93	24	35,5	939,7	22,8	21,82	1,07	11,94
11	14,66	2,01	15	43,4	940,8	21,6	33,02	0,41	12,63
12	14,8	2,06	12	46,2	940,5	21,6	33,02	0,41	12,63
13	14,93	1,96	15	48,3	940,5	21,6	33,02	0,41	12,63
14	16,06	1,61	13	36,3	934,6	31,9	23,45	0,92	8,57
15	16,21	1,53	13	36,6	933,8	31,9	23,45	0,92	8,57
16	16,33	1,33	13	36,6	934	31,9	23,45	0,92	8,57
17	16,95	1,71	21	32	934,2	40,9	23,49	0,96	10,01
18	17,06	1,71	23	31,1	934,2	40,9	23,49	0,96	10,01
19	17,18	1,77	26	30,8	934,2	40,9	23,49	0,96	10,01
20	14,2	0,96	13	45,3	932,4	46,2	25,735	0,96	10,63
21	14,3	0,89	13	45,4	932,4	47,2	25,735	0,96	10,63
22	14,41	1,01	12	46,9	932,2	48,2	25,735	0,96	10,63
23	14,85	1,61	11	47,5	931,5	29,5	35,295	0,86	13,63
24	14,95	1,68	12	46,9	931,9	30,5	35,295	0,86	13,63
25	15,1	1,57	12	46,2	931,9	31,5	35,295	0,86	13,63
26	15,21	1,15	14	44,3	931,9	33,2	29,14	0,73	10,63
27	15,45	1,06	11	47,4	931,9	35,2	29,14	0,73	10,63

Measure	Schedule	Issue	Umi. Air (°C)	Air temp. Air (°C)	P Atm. (hPa)	Humid. Soil (%)	Soil Temp. Soil (°C)	Cond. Tèrni (W½Γ^{-1} K)$^{-1}$	C/N
28	15,53	0,98	15	48	931,9	36,2	29,14	0,73	10,63
29	15,95	1,61	13	45	932,4	32,3	27,49	0,57	18,81
30	16,05	1,60	13	40,4	931,8	33,3	27,49	0,57	18,81
31	16,26	1,61	17	37,1	932	35,3	27,49	0,57	18,81
32	16,36	1,60	15	37,3	932	36,3	27,49	0,57	18,81
33	7,25	0,64	64	23,9	939	28,5	22,55	0,41	9,51
34	7,5	0,68	54	26,3	939,5	28,5	22,55	0,41	9,51
35	7,66	0,65	48	28	939	28,5	22,55	0,41	9,51
36	8	0,68	45	30,4	940,4	28,5	22,55	0,41	9,51
37	8,23	0,93	40	34	940,4	26,5	23,94	0,49	9,95
38	8,51	0,93	28	40,7	940,3	26,5	23,94	0,49	9,95
39	8,61	0,86	14	44,3	940	26,5	23,94	0,49	9,95
40	8,76	0,86	25	41,5	940	26,5	23,94	0,49	9,95
41	8,98	0,61	22	42,4	940,2	25,2	23,32	0,34	10,69
42	9,11	0,69	29	39,3	940,5	25,2	23,32	0,34	10,69
43	9,26	0,64	25	40,5	940,5	25,2	23,32	0,34	10,69
44	9,51	0,58	23,2	37,3	940,6	25,2	23,32	0,34	10,69
45	9,6	0,51	33	36,7	940,6	25,2	23,32	0,34	10,69
46	9,76	0,75	31	37,9	940,4	30,2	25,1	0,80	9,68
47	9,88	0,98	26	41,6	940,4	30,2	25,1	0,80	9,68
48	9,98	0,93	14	44,8	940,4	30,2	25,1	0,80	9,68
49	10,21	0,86	11	47,3	940	30,2	25,1	0,80	9,68
50	13,33	3,86	53	28	940,4	53,8	23,02	0,299	8,19

51	13,45	2,54	56	28	940,4	53,8	23,02	0,299	8,19
52	13,58	2,38	64	29,5	940,3	53,8	23,02	0,299	8,19
53	13,7	2,30	54	30,1	940,3	53,8	23,02	0,299	8,19
54	13,86	2,08	54	29,5	940,2	53,8	23,02	0,299	8,19
55	14,13	1,59	47	30,5	940,3	47,1	23,43	0,289	9,66*
Measure	Schedule	Issue	Umi. Air (°C)	Air temp. Air (° C)	P Atm. (hPa)	Humid. Soil (%)	Soil Temp. Soil (° C)	Cond. Tèrni. (W*nr^{-1} K)$^{-1}$	C/N
56	14,31	1,68	49	30,1	940,2	47,1	23,43	0,289	9,66*
57	14,56	1,56	56	29,5	940,3	47,1	23,43	0,289	9,66*
58	14,65	1,55	54	29,3	940,1	47,1	23,43	0,289	9,66*
59	14,8	1,56	66	28,6	939,9	47,1	23,43	0,289	9,66*
60	15,33	1,47	57	28,7	940	50,5	23,3	0,449	8,67
61	15,13	1,03	62	28,7	939,7	50,5	23,3	0,449	8,67
62	15,3	1,11	64	28,4	939,6	50,5	23,3	0,449	8,67
63	15,41	0,92	66	28,1	939,6	50,5	23,3	0,449	8,67
64	15,6	1,19	62	28,3	939,4	50,5	23,3	0,449	8,67
65	15,88	1,32	61	28,7	939,5	48,8	22,95	0,32	8,42
66	16,01	1,09	66	28,7	939,5	48,8	22,95	0,32	8,42
67	16,11	1,25	65	28,2	939,3	48,8	22,95	0,32	8,42
68	16,25	1,39	58	27,9	939,2	48,8	22,95	0,32	8,42
69	14	0,85	47	27	940,2	43,8	21,42*	0,48*	11,39
70	14,01	0,76	49	27	940,3	43,8	21,42*	0,48*	11,39
71	14,03	0,61	48	27,1	940,2	43,8	21,42*	0,48*	11,39
72	7,2	0,89	48	27,2	940,3	43,8	21,42*	0,48*	11,39
73	15	3,04	80	26	939,5	65,6	22,51	0,58	10,61
74	15,25	2,92	78	26,2	939,5	65,6	22,51	0,58	10,61
75	15,36	2,76	78	26,5	939,3	65,6	22,51	0,58	10,61
76	15,5	1,97	77	26	939,2	65,6	22,51	0,58	10,61
77	15,66	1,75	70	26	939,7	60,6	22,48	0,55	10,81
78	16	2,57	72	25,8	939,6	60,6	22,48	0,55	10,81
79	16,01	1,23	70	25,8	939,6	60,6	22,48	0,55	10,81
80	16,31	3,35	68	26	939,4	60,6	22,48	0,55	10,81
81	16,5	2,73	68	26	939,5	60,6	22,48	0,55	10,81
82	16,75	2,07	65	25,5	940,2	53,8	22,85	0,72	10,46
83	17	2,57	60	25,5	940,3	53,8	22,85	0,72	10,46
Measure	Schedule	Issue	Umi. Air (°C)	Air temp. Air (°C)	P Atm. (hPa)	Humid. Soil (%)	Soil Temp. Soil (°C)	Thermal Cond (W*m^{-1} K)$^{-1}$	C/N
56	14,31	1,68	49	30,1	940,2	47,1	23,43	0,289	9,66*
57	14,56	1,56	56	29,5	940,3	47,1	23,43	0,289	9,66*
58	14,65	1,55	54	29,3	940,1	47,1	23,43	0,289	9,66*
59	14,8	1,56	66	28,6	939,9	47,1	23,43	0,289	9,66*
60	15,33	1,47	57	28,7	940	50,5	23,3	0,449	8,67
61	15,13	1,03	62	28,7	939,7	50,5	23,3	0,449	8,67
62	15,3	1,11	64	28,4	939,6	50,5	23,3	0,449	8,67
63	15,41	0,92	66	28,1	939,6	50,5	23,3	0,449	8,67
64	15,6	1,19	62	28,3	939,4	50,5	23,3	0,449	8,67
65	15,88	1,32	61	28,7	939,5	48,8	22,95	0,32	8,42
66	16,01	1,09	66	28,7	939,5	48,8	22,95	0,32	8,42
67	16,11	1,25	65	28,2	939,3	48,8	22,95	0,32	8,42
68	16,25	1,39	58	27,9	939,2	48,8	22,95	0,32	8,42
69	14	0,85	47	27	940,2	43,8	21,42*	0,48*	11,39
70	14,01	0,76	49	27	940,3	43,8	21,42*	0,48*	11,39
71	14,03	0,61	48	27,1	940,2	43,8	21,42*	0,48*	11,39
72	7,2	0,89	48	27,2	940,3	43,8	21,42*	0,48*	11,39
73	15	3,04	80	26	939,5	65,6	22,51	0,58	10,61
74	15,25	2,92	78	26,2	939,5	65,6	22,51	0,58	10,61
75	15,36	2,76	78	26,5	939,3	65,6	22,51	0,58	10,61
76	15,5	1,97	77	26	939,2	65,6	22,51	0,58	10,61
77	15,66	1,75	70	26	939,7	60,6	22,48	0,55	10,81
78	16	2,57	72	25,8	939,6	60,6	22,48	0,55	10,81

Measure	Schedule	Issue	Umi. Air (°C)	Air temp. Air (°C)	P Atm. (hPa)	Humid. Soil (%)	Soil Temp. Soil (°C)	Thermal Cond. (W*m^{-1} K)$^{-1}$	C/N
79	16,01	1,23	70	25,8	939,6	60,6	22,48	0,55	10,81
80	16,31	3,35	68	26	939,4	60,6	22,48	0,55	10,81
81	16,5	2,73	68	26	939,5	60,6	22,48	0,55	10,81
82	16,75	2,07	65	25,5	940,2	53,8	22,85	0,72	10,46
83	17	2,57	60	25,5	940,3	53,8	22,85	0,72	10,46
84	17,25	2,86	60	25,3	939,5	53,8	22,85	0,72	10,46
85	17,5	3,02	60	25,3	939,5	53,8	22,85	0,72	10,46
86	14,61	1,59	81	25,2	945,4	57,2	21,17	0,66	8,77
87	14,75	1,95	84	25,2	945,4	57,2	21,17	0,66	8,77
88	14,86	1,99	81	25,4	945,2	57,2	21,17	0,66	8,77
89	15,16	1,98	80	25,2	945,1	57,2	21,17	0,66	8,77
90	10,38	1,64	86	23,2	945,9	57,2	21,42*	0,48*	9,66
91	10,66	1,73	88	23,3	945,6	53,8	21,42*	0,48*	9,66
92	10,91	2,59	89	23,5	945,3	63,9	21,42*	0,48*	9,66
93	11,08	2,14	89	24,2	945,2	50,5	21,42*	0,48*	9,66
94	11,33	2,49	88	24,7	945,0	67,3	21,42*	0,48*	9,66
95	9,83	2,10	89	19,4	948,6	39,9	18	0,48*	9,66
96	10,08	2,03	88	19,6	948,8	39,9	18	0,48*	9,66
97	10,25	1,67	92	18,9	948,5	34,0	18	0,48*	9,66
98	9,58	2,40	77	21	949,3	70,0	18	0,48*	9,66

ANNEX 02 - Multiple linear regression results

Area	Measured Issue	Observed Emission	Calculated Emission Multiple Linear Regression	White-Hubber	Robusta
TALK 15	1	1,08	0,84	0,84	0,95
	2	2,29	2,20	2,20	2,19
	3	2,23	2,04	2,04	2,06
	4	2,01	1,89	1,89	1,90
	5	2,10	1,78	1,78	1,82
	6	2,17	1,83	1,83	1,85
	7	2,28	1,74	1,74	1,78
	8	1,79	1,46	1,46	1,68
	9	1,84	1,36	1,36	1,59
	10	1,93	1,44	1,44	1,67
	11	2,01	1,87	1,87	1,88
	12	2,06	1,72	1,72	1,74
	13	1,96	1,53	1,53	1,60
	14	1,61	1,37	1,37	1,38
	15	1,53	1,26	1,26	1,28
	16	1,33	1,29	1,29	1,31
	17	1,71	1,91	1,91	1,89
	18	1,71	1,94	1,94	1,92
	19	1,77	1,91	1,91	1,90
	20	0,96	1,35	1,35	1,26
	21	0,89	1,38	1,38	1,28
	22	1,01	1,32	1,32	1,22
	23	1,61	1,62	1,62	1,58
	24	1,68	1,72	1,72	1,67
	25	1,57	1,80	1,80	1,75
	26	1,15	1,06	1,06	1
	27	1,06	0,99	0,99	0,94

Area	Measured Issue	Observed Emission	Calculated Emission Multiple Linear Regression	White-Hubber	Robusta
	28	0,98	0,92	0,92	0,88
	29	1,61	1,24	1,24	1,42
	30	1,60	1,50	1,50	1,63
	31	1,61	1,74	1,74	1,84
	32	1,60	1,80	1,80	1,88
	33	0,64	0,74	0,74	0,69
	34	0,68	0,84	0,84	0,77
	35	0,65	0,79	0,79	0,71
	36	0,68	0,86	0,86	0,81
	37	0,93	0,90	0,90	0,88
	38	0,93	0,69	0,69	0,69
	39	0,86	0,70	0,70	0,66
	40	0,86	0,68	0,68	0,66
	41	0,61	0,51	0,51	0,51
Area	Measured Issue	Observed Emission	Calculated Emission Multiple Linear Regression	White-Hubber	Robusta
TALHÃO 15	42	0,69	0,61	0,61	0,63
	43	0,64	0,62	0,62	0,63
	44	0,58	0,88	0,88	0,85
	45	0,51	0,74	0,74	0,75
	46	0,75	1,39	1,39	1,39
	47	0,98	1,25	1,25	1,27
	48	0,93	1,27	1,27	1,26
	49	0,86	1,13	1,13	1,14
TABLE 23	50	3,86	1,82	1,82	1,57
	51	2,54	1,77	1,77	1,54
	52	2,38	1,52	1,52	1,34
	53	2,30	1,67	1,67	1,46
	54	2,08	1,70	1,70	1,49
	55	1,59	1,68	1,68	1,54
	56	1,68	1,66	1,66	1,53
	57	1,56	1,59	1,59	1,49
	58	1,55	1,62	1,62	1,51
	59	1,56	1,42	1,42	1,37
	60	1,47	1,81	1,81	1,70
	61	1,03	1,67	1,67	1,58
	62	1,11	1,65	1,65	1,57
	63	0,92	1,63	1,63	1,56
	64	1,19	1,68	1,68	1,60
	65	1,32	1,47	1,47	1,39
	66	1,09	1,38	1,38	1,33
	67	1,25	1,42	1,42	1,35
	68	1,39	1,56	1,56	1,46
	69	0,85	1,84	1,84	1,82
	70	0,76	1,81	1,81	1,80
	71	0,61	1,81	1,81	1,80
	72	0,89	1,56	1,56	1,39
	73	3,04	2,15	2,15	2,05
	74	2,92	2,18	2,18	2,08
	75	2,76	2,15	2,15	2,05

Area	Measured Issue	Observed Emission	Calculated Emission Multiple Linear Regression	White-Hubber	Robusta
	76	1,97	2,19	2,19	2,09
	77	1,75	2,19	2,19	2,11
	78	2,57	2,17	2,17	2,10
	79	1,23	2,21	2,21	2,13
	80	3,35	2,22	2,22	2,14
	81	2,73	2,24	2,24	2,17
	82	2,07	2,35	2,35	2,36
	83	2,57	2,47	2,47	2,45
TALK 23	84	2,86	2,40	2,40	2,39
	85	3,02	2,41	2,41	2,40
	86	1,59	2,36	2,36	2,38
	87	1,95	2,31	2,31	2,34
	88	1,99	2,34	2,34	2,36
	89	1,98	2,37	2,37	2,39
	90	1,64	2,21	2,21	2,11
	91	1,73	2,02	2,02	1,98
	92	2,59	2,32	2,32	2,18
	93	2,14	1,80	1,80	1,82
	94	2,49	2,36	2,36	2,20
	95	2,10	1,73	1,73	1,87
	96	2,03	1,77	1,77	1,91
	97	1,67	1,51	1,51	1,73
	98	2,40	2,97	2,97	2,75